PRAISE FOR *The Triumph of Seeds*

"[An] engaging book. . . . What makes *The Triumph of Seeds* more than a routine pop botany book is the way Mr. Hanson teases out the resonances between the ways that plants and humans use seeds. . . . [A] lively and intelligent book."
—Richard Mabey, *Wall Street Journal*

"The genius of Hanson's fascinating, inspiring, and entertaining book stems from the fact that it is not about how all kinds of things grow from seeds; it is about the seeds themselves. Hanson . . . takes one of the least-impressive-looking natural objects and reveals a life of elegance and wonder. . . . Although he is a storyteller by nature, he also charms us with an infectious enthusiasm. The reader feels that Hanson cannot wait to tell us what comes next. . . . Like all good writers, he understands narrative—that a book, at its best, is a story, and that this one is built by spinning stories within stories. They are fun, sometimes they are funny, and they are always fascinating and readable. . . . [An] engaging book."
—Mark Kurlansky, *New York Times Book Review*

"With light, engaging prose Hanson shows how the little spheroids we tip out of a packet are in fact supremely elegant genetic time capsules. *The Triumph of Seeds* takes you past the casing into the extraordinary inner workings of objects without which our landscapes, dinner plates, and gardens would be unrecognizable. You will never be able to look at an orange pip or a sunflower seed in the same way again."
—*New Scientist*

"A rip-roaring read."

—Robert Krulwich, *National Geographic*'s Curiously Krulwich blog

"An intriguing look at the acorns that grow into oaks, the orchid beans that flavor vanilla extract, and other ordinary seeds that affect the world, often in extraordinary ways. . . . [*The Triumph of Seeds*] is a mix of lively stories, adventure, natural history, botany, and ecology. . . . Hanson's book isn't a 'how-to,' but it is a 'don't miss' for naturalists, from amateurs to experts, or for anyone who enjoys growing plants from seeds."

—HGTVGardens.com

"Thor Hanson has taken the history and science of these little marvels and drawn out a fascinating account of seed culture. We should not forget the importance of seeds in the liquor cabinet, as well. From rye whiskey, to wheat vodka, to barley beer, it would be a lot harder to get drunk without our friends in the grain world."

—*Home Wet Bar* blog

"[Hanson's] luck for finding then writing about the magic in something common continues with *The Triumph of Seeds*."

—*Seattle Times*

"Lest you get the impression that Hanson's book is all academic grit and gruel, be advised that he has thoroughly leavened his narrative with odd facts and fascinating digressions."

—*Natural History*

"This is a charming book, inspired by Hanson's forays into seed identification and dispersal with his young, seed-obsessed son. . . . Hanson's twist of looking at human interactions with plants in their embryonic stage is new. . . . *The Triumph of Seeds* will engender thoughtful consideration of our joint future."

—*Nature*

"A delight. Composed in charming and lively prose, the book introduces readers to a variety of quirky figures—biologists, farmers, archaeologists, and everyday gardeners—who have something profound to say about a seemingly mundane topic: those little kernels that, against tremendous odds, have managed to take root all around us. . . . *The Triumph of Seeds* is a remarkable, gentle, and refreshing piece of work that draws readers further into the wide arms of the world and makes them grateful for it."

—*BookPage*

"Conservation biologist Hanson's new book showcases an even more approachable style than his 2011 *Feathers*. Using a personalized viewpoint derived from his backyard lab and dissertation research in Costa Rica with the almendro tree, as well as visits with specialists worldwide, he describes how seeds nourish, unite, endure, defend, and travel."

—*Library Journal*

"Fast and fascinating prose. . . . Hanson, who has also chronicled feathers and gorillas, is a conservation biologist and Guggenheim fellow, and an ace dot-connector: he can draw a line between all the grain panics and crises and the tiny, miraculous structure of the seeds themselves, because he dives deeply into botany, economy and history. Also, he's just plain fun."

—*Denver Post*

"[Hanson is] jocular and entertaining in his dispensing of remarkable facts about these little vessels of life-to-be. . . . From high-tech, high-security seed banks bracing for climate change to the story of the gum extracted from guar seeds that is used in everything from ice cream to fracking, this upbeat and mind-expanding celebration of the might of seeds is popular science writing at its finest."

—*Booklist*, starred review

"Hanson's writing is lively, inquisitive, and knowledgeable. He draws on his own knowledge and that of a wide field of experts, writing a clear, comprehensible book that covers a wide range of topics."

—*Fangirl Nation*

"A delightful account of the origins, physiologies, and human uses of a vast variety of objects that plants employ to make more plants. . . . A fine addition to the single-issue science genre."

—*Kirkus Reviews*

"Hanson writes in that breezy, enthused, confident way of good American science writers, scattering stories and analogies like dandelion seed-puffs. . . . [*The Triumph of Seeds*] is a good example of extrovert nature writing, weaving together biology, human history, and 'juicy seed lore.'"

—*British Wildlife* (UK)

"For the past fifty-seven years I have relied on seeds for food and, throughout much of my horticultural career, for earning a living. This new book has re-excited my fascination [with] these wonderful living structures. . . . [I]nformative, yet readable. . . . [A] fascinating book."

—Chris Allen, *The Gardening Times* (UK)

"Who knew that seeds could be so thrilling and dangerous? Thor Hanson is a lively storyteller, a lyrical writer, and a quick wit. *The Triumph of Seeds* is more than an engrossing work of natural history. It's a compelling and highly entertaining journey, populated by scientists and historians, criminals and explorers, aviators and futurists. Following Hanson's global voyage is the best sort of armchair travel, because it is filled with wonder, poetry, and discovery."

—Amy Stewart, author of *The Drunken Botanist:*
   The Plants That Create the World's Great Drinks,
   a *New York Times* Bestseller

"This beautifully written book is a magnificent read. Every page is full of surprises and illuminating insights, illustrating the fascinating evolution of seeds, and their extraordinary impact on humans, past and present. A master storyteller, Hanson has created a first-rate natural history. When you reach the end of this page-turner, you will wish there were more . . . and you will never look at seeds in the same way."

—Eric Jay Dolin, author of *Leviathan: The History of Whaling in America* and *When America Met China: An Exotic History of Tea, Drugs, and Money in the Age of Sail*

"As he did in his phenomenal *Feathers*, Thor Hanson brings us the incredible world of seeds in a package as graceful and elegant as they are themselves, gift-wrapped in utterly seductive stories. I cannot recall a book I was so eager to finish, that I might begin it again."

—Robert Michael Pyle, author of *Wintergreen* and *Mariposa Road*

"If you eat seeds of any kind, you must read this book! Ecologist Hanson gives us a rich Darwinian view of how seeds came to be the most important nutritional resource for human as well as older than human species. He is at his best when we are in the field with him, learning like detectives the 'whodunnits' of seed dispersal. You will never look a seed in the eye again without thanking Thor."

—Gary Paul Nabhan, Franciscan brother and author of *Enduring Seeds* and *Food, Genes, and Culture*

"Thor has done it again. In a page-turner, he tells the stories of seeds, their ecology, evolution, and histories and why each of us every day depends on, relies on, delights in, or suffers from seeds. This book will change the way you think about your coffee, your chocolate, or even just the weed growing stubbornly (from a seed) out of the crack in the sidewalk. Seeds are everywhere—a reality that you will never forget again after reading this book."

—Rob Dunn, author of *The Wild Life of Our Bodies*

THE TRIUMPH OF SEEDS

ALSO BY THOR HANSON

Feathers: The Evolution of a Natural Miracle
The Impenetrable Forest: Gorilla Years in Uganda

The
Triumph of

SEEDS

*How Grains, Nuts, Kernels, Pulses, & Pips
Conquered the Plant Kingdom
and Shaped Human History*

Thor
Hanson

BASIC BOOKS
A Member of the Perseus Books Group
New York

Hardcover first published in 2015 by Basic Books,
A Member of the Perseus Books Group

Paperback first published in 2016 by Basic Books

Books published by Basic Books are available at special discounts for bulk purchases in the United States by corporations, institutions, and other organizations. For more information, please contact the Special Markets Department at the Perseus Books Group, 2300 Chestnut Street, Suite 200, Philadelphia, PA 19103, or call (800) 810-4145, ext. 5000, or e-mail special.markets@perseusbooks.com.

Designed by Trish Wilkinson
Set in 10.5 point Goudy Oldstyle Std

The Library of Congress has cataloged the original edition as follows:
Hanson, Thor, author.
    The triumph of seeds : how grains, nuts, kernels, pulses, and pips, conquered the plant kingdom and shaped human history / Thor Hanson.
        pages   cm
    Includes bibliographical references and index.
    ISBN 978-0-465-05599-9 (hardcover : alk. paper)—ISBN 978-0-465-04872-4 (e-book) 1. Seeds. I. Title.
QK661.H36 2015
581.4'67—dc23

2014047078

ISBN 978-0-465-09740-1 (paperback)

10 9 8 7 6 5 4 3 2

For Eliza and Noah

Contents

Seeds Defend _____

Seeds Travel _____

Author's Note

Throughout this book I have chosen to stick with a functional definition of seeds, acknowledging that in some cases the seed-like part of a plant might also include tissues derived from the fruit (e.g., the shell of a nut). The text includes only common names of plants, but a complete list of all Latin binomials is included in Appendix A. I've tried to keep botanical jargon to an absolute minimum, or explain it with context, but have also compiled a short glossary (also at the back of the book). Finally, I encourage readers not to neglect the notes for each chapter. They include a wealth of juicy seed lore that couldn't be squeezed into the narrative, but was too good to leave out entirely.

Acknowledgments

In writing this book I have relied on the help and patience of a huge range of generous people. Here, in no particular order, is a partial list of those who have given assistance along the way— granting interviews, loaning books and papers, answering questions, and even pitching in with some timely babysitting: Carol and Jerry Baskin, Christina Walters, Robert Haggerty, Bill DiMichele, Fred Johnson, John Deutch, Derek Bewley, Patrick Kirby, Richard Wrangham, Sam White, Michael Black, Chris Looney, Ole J. Benedictow, Micaela Colley, Amy Grondin, John Navazio, Matthew Dillon, Sarah Shallon, Elaine Solowey, Hugh Pritchard, Howard Falcon-Lang, Matt Stimson, Scott Elrick, Stanislav Opluštil, Bob Sievers, Phil Cox, Robert Druzinsky, Greg Adler, David Strait, Judy Chupasko, Diane Ott Whealy, Sophie Rouys, Pam Stuller, Noelle Machnicki, Chelsey Walker-Watson, Brandon Paul Weaver, Hiroshi Ashihara, Jeri Wright, Ronald Griffiths, Chifumi Nagai, Steve Meredith, David Newman, Richard Cummings, Giovanni Giustina, Jason Werden, Erin Braybrook, the International Spy Museum, Valéria Forni Martins, Mark Stout, Al and Nellie Habegger, Thomas Boghardt, Ira Pastan, Kirsten Gallaher, Uno Eliasson, Jonathan Wendel, Duncan Porter, Charles Moseley, Boyd Pratt, Bella French, Paul Hanson, Aaron Burmeister, Nason and Erica Hamlin,

John Dickie, Suzanne Olive, Amy Stewart, Derek and Susan Arndt, Kathleen Ballard, and Chris Weaver.

I am indebted to the John Simon Guggenheim Memorial Foundation for granting me a fellowship specifically in support of the writing of this book. The Leon Levy Foundation also contributed generously to that fellowship.

Special thanks for help with research go to the University of Idaho Library and to the San Juan Island Library, with particular appreciation for the patience and diligence of Interlibrary Loan Coordinator Heidi Lewis.

I'm grateful for the talents and enthusiasm of my agent, Laura Blake Peterson, and all her colleagues at Curtis Brown, and it has once again been a true pleasure to work with TJ Kelleher and the superlative team at Basic Books and Perseus, including (but not limited to) Sandra Beris, Cassie Nelson, Clay Farr, Michele Jacob, Trish Wilkinson, Nicole Jarvis, and Nicole Caputo.

Finally, none of this would be possible, or particularly enjoyable, without the love and support of my friends and family.

THE TRIUMPH OF SEEDS

Preface
"Heed!"

Sir, I can nothing say,
But that I am your most obedient servant.

— William Shakespeare,
All's Well That Ends Well (c. 1605)

Charles Darwin traveled with the HMS *Beagle* for five years, devoted eight years to the anatomy of barnacles, and spent most of his adult life ruminating on the implications of natural selection. The famed naturalist-monk Gregor Mendel hand-pollinated 10,000 pea plants over the course of eight Moravian springtimes, before finally writing up his thoughts on inheritance. At Olduvai Gorge, two generations of the Leaky family sifted through sand and rock for decades to piece together a handful of critical fossils. Unraveling evolutionary mysteries is generally hard work, the stuff of long careers spent in careful thought and observation. But some stories are obvious, crystal clear from the very beginning. Anyone familiar with children, for example, understands the origin of punctuation. It started with the exclamation point.

Nothing comes more naturally to a toddler than emphatic, imperative verbs. In fact, any word can be transformed into a command with the right inflection—a gleeful, insistent shout accented from a seemingly bottomless quiver of exclamation points. Whatever nuances of speech and prose might be gained by the use of comma, period, or semicolon clearly developed later. The exclamation point is innate.

Our son, Noah, is a good example. He began his verbal career with many of the expected phrases, from "Move!" and "More!" to the always-popular "No!" But his early vocabulary also reflected a more unusual interest: Noah was obsessed with seeds. Neither Eliza nor I can remember exactly when this passion began; it just seemed that he had always loved them. Whether speckling the skin of a strawberry, scooped from inside a squash, or chewed up in the rose hips he plucked from roadside shrubs, any seed that Noah encountered was worthy of attention and comment. In fact, determining which things had seeds, and which didn't, became one of the first ways he learned to order his world. Pinecone? Seeds. Tomato? Seeds. Apple, avocado, sesame bagel? All with seeds. Raccoon? No seeds.

With such conversations a regular occurrence in our household, it's no wonder that seeds were on my short list when it was time to settle on a new book idea. What might have tipped the balance was Noah's pronunciation, which added a certain imperative to his botanical observations. Sibilance did not come easily to his young tongue, but instead of lisping he chose to replace 's' sounds with a hard 'h.' The result was a barrage of double commands—every time he disassembled some unsuspecting piece of fruit he would raise the seeds in my direction and shout, "HEED!" Day after day, this scene repeated itself until I eventually got the message: I heeded the seeds. After all, little Noah had already pretty much taken over the rest of our lives. Why not put him in charge of career decisions, too?

Fortunately, he assigned me a topic dear to my heart, a book that I'd wanted to write for years. As a doctoral student, my research included studies of seed dispersal and seed predation in huge tropical rainforest trees. I learned how vital those seeds were, not just to the trees but to the bats and monkeys that dispersed them; the parrots, rodents, and peccaries that ate them; the jaguars that hunted the peccaries; and so on. Researching seeds enriched my understanding of biology, but it also taught me how their influence reaches far beyond the edge of forest or field; seeds are vital everywhere. They transcend that imaginary boundary we erect between the natural world and the human world, appearing so regularly in our daily lives, in so many forms, that we hardly recognize how utterly dependent we are upon them. Telling their story reminds us of our fundamental connections to nature—to plants, animals, soil, seasons, and the process of evolution itself. And in an age where, for the first time, more than half the human population lives in cities, reaffirming those links has never been more important.

Before this tale travels even another paragraph, however, I need to insert two caveats. The first is an important clarification that will preserve good relations with my many friends in marine biology. In the 1962 film *Mutiny on the Bounty*, there is a memorable scene where the rebellious sailors set Captain Bligh adrift in a longboat and then immediately toss every one of his hated breadfruit seedlings overboard. (Bligh had been giving the plants regular doses of fresh water even after the crew's rations ran low.) As the little trees go over the side of the ship, the camera pans back to show them trailing in the *Bounty*'s wake: a handful of pitiful green motes on a vast, calm sea. Their prospects look dim, making an important point about the limitations of the seed strategy. While seed plants may triumph on dry land, different rules apply to the nearly three-quarters of the planet covered by oceans. There, algae and tiny phytoplankton hold sway, limiting their seed-bearing cousins to a few shallow-water varieties, the occasional bobbing coconut, and things cast off by sailors. Seeds evolved on terra firma, where their many remarkable traits have shaped the course of natural and human history. But it's good to keep in mind that on the open ocean, they're still a novelty act.

The second caveat acknowledges an area of seed controversy that lies beyond the scope and aim of this book. In graduate school, my curriculum included a one-credit seminar intended to familiarize students with the equipment used in a genetics laboratory. We met in the evenings once a week, donned white lab coats, and spent a couple of hours practicing with various tubes and pipes and whirring, beeping machines. As a simple exercise, the instructor showed us how to splice our own DNA into that of a bacterial cell. As the bacterial colony then divided and grew, our DNA would be copied ad infinitum, a basic form of cloning. Though of course we only used a tiny fragment of DNA and the results were crude, I distinctly remember thinking, *"I shouldn't be able to clone myself in a one-credit class."*

The advent of relatively straightforward techniques for genetic manipulation has ushered in a new era for plants and their seeds. Familiar crops, from corn and soybeans to lettuce and tomatoes, have been experimentally altered with genes borrowed from arctic fish (for frost resistance), soil bacteria (to make their own pesticide), and even *Homo sapiens* (to produce human insulin). Seeds can now be patented as intellectual property, and designed to include *terminator genes* that prevent the ancient practice of saving seed for future plantings. Genetic modification is a pivotal new technology, but it will be addressed only briefly in these pages. Instead, this book explores why we care so much in the first place. When modern genetics has also given us featherless chickens, glow-in-the-dark cats, and goats that produce spider silk, why is it that seeds are the focal point of the debate? Why do polls consistently find people more comfortable with the idea of changing their own genome, or the genomes of their children (for medical purposes), than they are with the notion of altering the genes in seeds?

The answers to these questions lie in a story that stretches back millions of years, wonderfully entwining the history of seeds with the history of our own species and culture. The challenge for me in writing this book lay not in filling it, but in deciding what material to include and what must be left by the side of the road. (For additional anecdotes and information, be sure to read the notes for each chapter. They are the only place in the book, for example, where you will hear about gomphotheres, slippery water, or the piper's maggot.) Along the way we will meet fascinating plants and animals as well as many people who have made seeds a part of their own stories, from scientists and farmers to gardeners, merchants, explorers, and chefs. If I have done my job right, you will see in the end what I have come to know, and what Noah apparently realized from the start: seeds are a marvel, worthy of our study, praise, wonder, and any number of exclamations points. (!)

The Fierce Energy

Think of the fierce energy concentrated in an acorn! You bury
it in the ground, and it explodes into a giant oak! Bury a sheep,
and nothing happens but decay.

—George Bernard Shaw,
The Vegetarian Diet According to Shaw (1918)

I put the hammer down and peered at the seed. Not a scratch. Its
dark surface looked just as smooth and perfect as it had when
I'd found it on the floor of the rainforest. There, lying in the mud
and mulch, surrounded by the sounds of water dripping and the
constant chirr of insects, the seed had looked ready to burst open
with the promise of bud swell, roots, and leafy green. Now, under
the hum of fluorescent lights in my office, the damn thing seemed
indestructible.

I picked up the seed and it settled neatly in my palm—a little
larger than a walnut but flatter and dark, its heavy shell as hard as
tempered steel. A thick seam ran lengthwise around the edge, but
no amount of prodding and prying with a screwdriver had opened
it so much as a crack. Vigorous squeezing with a long-handled pipe
wrench hadn't proved any better, and now hammer blows appeared
to be useless. Obviously, I needed something heavier.

My university office took up a corner of the old Forestry Department herbarium, a largely forgotten place where dried plant collections lined the walls in dusty metal cabinets. Once a week, a group of retired faculty members gathered there for coffee and bagels, reminiscing about research trips, favorite trees, and departmental infighting from decades past. My desk, too, dated from an earlier era, a time when people built office furniture from welded steel, chrome, and double-weight Formica. It was large enough for a fleet of mimeograph machines and teletypes, and hefty enough to withstand the shockwaves of a nuclear attack.

Placing the seed close beside one of its hulking feet, I heaved the desk upward and let fly. It dropped with a resounding crash, launching the seed sideways to ricochet off the wall and skitter out of sight under a cabinet. When I retrieved it, the seed's dark sides looked completely unscathed. So I tried again—*crash!*—and again—*crash!*—my frustration mounting with every failed attempt. Finally, I crouched down, pinned the seed between desk leg and wall, and started flailing at it wildly with the hammer.

My anger in that moment, however, didn't come close to that of the red-faced forestry professor who suddenly stormed into the room, shouting, "What the hell is going on in here? I'm trying to teach a class next door!"

Clearly, I needed a quieter seed-breaking method. Particularly since it wasn't just one seed that I needed to open. Two crates in the closet held hundreds of them, not to mention more than 2,000 leaves and bits of bark, each one painstakingly gathered over months of fieldwork in the forests of Costa Rica and Nicaragua. Transforming those samples into data would make up the bulk of my doctoral dissertation. Or not, as things were going.

Eventually, I discovered that a stout blow with a mallet and rock chisel would do the job nicely, but my struggle to open that first seed taught me an important evolutionary lesson. Why, I asked myself, would a seed's shell be so impossibly hard to split? Wasn't the whole point of a seed to throw itself wide and let the young plant spring forth? Surely, that thick husk hadn't evolved merely to thwart a hapless graduate student. The answer, of course, was as fundamental as a broody hen guarding her clutch of eggs, or a lioness defending her cubs. To the tree I was studying, the next generation meant everything, an evolutionary imperative worth any investment of energy and adaptive creativity. And in the history of plants, no single event has ensured the protection, dispersal, and establishment of their progeny more than the invention of seeds.

In business, people mark the ultimate success of a product by its brand recognition and universal availability. When I lived in a mud-walled hut in Uganda, four hours from a paved road, on the edge of a jungle called the Impenetrable Forest, I could still buy a bottle of Coca-Cola within a five-minute walk from my front door. Marketing executives fantasize about that kind of ubiquity, and in the natural world, seeds have it. From tropical rainforests to alpine meadows and arctic tundra, seed plants dominate landscapes and define entire ecosystems. A forest, after all, is named for its trees and not for the monkeys or birds that leap and flutter within it. And everyone knows to call the famed Serengeti a *grass*land—not a zebra-land with grass. When we pause to examine the underpinnings of natural systems, time and again we find seeds, and the plants that bear them, playing the most vital roles.

While an ice-cold soda tastes pretty good on a tropical afternoon, the Coca-Cola analogy only goes so far in explaining the evolution of seeds. But it is true in one more respect: natural selection, like commerce, rewards a good product. The best adaptations spread through time and space, in turn spurring further innovation in a process Richard Dawkins aptly called, "The Greatest Show on Earth." Some traits become so widespread they seem axiomatic. Animal heads, for example, have two eyes, two ears, some kind of nose, and a mouth. Fish gills extract dissolved oxygen from water. Bacteria reproduce by splitting, and the wings of insects come in pairs. Even for biologists, it's easy to forget that these fundamentals were once brand new, clever novelties spun from the sheer persistence of evolution's trial and error. In the plant world, the idea of seeds ranks right alongside photosynthesis as one of our chief assumptions. Even children's literature takes the notion for granted. In Ruth Krauss's classic book *The Carrot Seed*, a silent little boy disregards all naysayers, patiently watering and weeding around his planting until at last a great carrot sprouts up, "just as the little boy had known it would."

Though famous for how its simple drawings transformed the genre of picture books, Krauss's story also tells us something profound about our relationship with nature. Even children know that the tiniest pip contains what George Bernard Shaw called "fierce energy"—the spark and all the instructions needed to build a carrot, an oak tree, wheat, mustard, sequoias, or any one of the estimated 353,000 other kinds of plants that use seeds to reproduce. The faith we place in that ability gives seeds a unique position in the history of the human endeavor. Without the act and anticipation of planting and harvest, there could be no agriculture as we know it, and our species would still be wandering in small bands of hunters, gatherers, and herdsmen. Indeed, some experts believe that *Homo sapiens* might never have evolved at all in a world that lacked seeds. More than perhaps any other natural objects, these small botanical marvels paved the way for modern civilization, their fascinating evolution and natural history shaping and reshaping our own.

We live in a world of seeds. From our morning coffee and bagel to the cotton in our clothes and the cup of cocoa we might drink before bed, seeds surround us all day long. They give us food and fuels, intoxicants and poisons, oils, dyes, fibers, and spices. Without seeds there would be no bread, no rice, no beans, corn, or nuts. They are quite literally the staff of life, the basis of diets, economies, and lifestyles around the globe. They anchor life in the wild, too: seed plants now make up more than 90 percent of our flora. They are so commonplace it's hard to imagine that for over 100 million years other types of plant life dominated the earth. Roll back the clock and we find seeds evolving as trivial players in a flora ruled by spores, where tree-like club mosses, horsetails, and ferns formed

vast forests that remain with us in the form of coal. From this humble beginning, the seed plants steadily gained advantage—first with
conifers, cycads, and ginkgos, and then in a great diversification of
flowering species—until now it is the spore bearers and algae that
watch from the sidelines. This dramatic triumph of seeds poses an
obvious question: Why are they so successful? What traits and habits
have allowed seeds, and the plants that bear them, to so thoroughly
transform our planet? The answers frame the narrative of this book
and reveal not only why seeds thrive in nature, but why they are so
vital to people.

Seeds Nourish. Seeds come pre-equipped with a baby plant's
first meal, everything needed to send forth incipient root, shoot,
and leaf. Anyone who has ever put sprouts on a sandwich takes this
fact for granted, but it was a critical step in the history of plants.
Concentrating that energy into a compact, portable package opened
up a huge range of evolutionary possibilities and helped seed plants
spread across the planet. For people, unlocking the energy contained in seeds paved the way for modern civilization. To this day,
the foundation of the human diet lies in co-opting seed food, stealing the nourishment designed for baby plants.

Seeds Unite. Before seeds, plant sex was pretty dull stuff. When
they did it at all, plants made sure the act was quick, out of sight,
and usually with themselves. Cloning and other asexual means were
common, and whatever sex happened rarely mixed genes in a predictable or thorough way. With the advent of seeds, plants suddenly
began breeding in the open air, dispersing pollen to egg in increasingly creative ways. It was a profound innovation: unite the genes
from two parents on the mother plant and package them into portable, ready-to-sprout offspring. Where spore plants interbred only
occasionally, seed plants mixed and remixed their genes constantly.
The evolutionary potential was enormous, and it's no coincidence
that Mendel solved the mystery of inheritance by a close examination of pea seeds. Science might still be waiting to understand genetics if that famed pea experiment had instead been "Mendel's Spores."

Seeds Endure. As any gardener knows, seeds stored through the winter months can be planted the following spring. In fact, many seeds require a cold spell, a fire, or even passage through an animal gut to trigger their germination. Some species persist in the soil for decades, sprouting only when the right combination of light, water, and nutrients makes conditions right for plant growth. This habit of dormancy sets seed plants apart from nearly all other life forms, allowing great specialization and diversification. For people, mastering the storage and manipulation of dormant seeds paved the way for agriculture and continues to determine the fate of nations.

Seeds Defend. Almost any organism will fight to protect its young, but plants equip their seeds with an astonishing and sometimes deadly assortment of defenses. From impenetrable husks and jagged spikes to the compounds that give us hot peppers, nutmeg, and allspice, not to mention poisons like arsenic and strychnine, seed defenses include some surprising (and surprisingly useful) adaptations. Exploring this topic illuminates a major evolutionary force in nature and shows how people have co-opted seed defense for their own ends, from the heat in Tabasco sauce to pharmaceuticals to the most beloved seed products of all, coffee and chocolate.

Seeds Travel. Whether tossed up by storm waves, spun on the wind, or packaged in the flesh of a fruit, seeds have found countless ways to get around. Their adaptations for travel have given them access to habitats spanning the globe, spurred diversity, and led people to some of the most essential and valuable products in history, from cotton and kapok to Velcro and apple pie.

This book is both an exploration and an invitation. Like seeds themselves, it began as something small, an interest that grew with my own curiosity, following the winding path that seeds have paved through evolution, natural history, and human culture. From the jungles and laboratories of my own research, and at the insistence of my seed-obsessed son, I plunged in and let the story unfold, guided by the gardeners, botanists, explorers, farmers, historians, and monks I met along the way—not to mention the wonderful plants themselves, and the menagerie of animals, birds, and insects that depend upon them. But for all the fascinating tales of seeds in nature, one of their hallmarks is that we don't have to look far to find them. Seeds are an integral part of our world too. So whether you prefer coffee and chocolate-chip cookies or mixed nuts, popcorn, pretzels, and a glass of beer, I invite you to sit down with your favorite seed-derived snacks, and let the journey begin.

Seeds Nourish

Oats, peas, beans and barley grow,
Oats, peas, beans and barley grow,
Can you or I or anyone know
How oats, peas, beans and barley grow?

First the farmer sows his seed,
Stands erect and takes his ease,
He stamps his foot and claps his hands,
And turns around to view his lands.

—Traditional folk song

CHAPTER ONE

Seed for a Day

*I have great faith in a seed. Convince me that you have a seed
there, and I am prepared to expect wonders.*

—Henry David Thoreau,
The Dispersion of Seeds (1860–1861)

When a pit viper strikes, physics tells us it can't lunge forward
farther than the length of its own body. The head and front
end are agile, but the tail of the beast stays put. Anyone who has
been struck at, however, knows that these snakes can fly through
the air like Zulu spears, or the daggers thrown in ninja movies. The
one coming at me darted up from a mat of dead leaves, launching
itself at my boot in a lightning blur of fangs and intent. I recognized
it as a fer-de-lance, a snake famed throughout Central America for
its unfortunate combination of strong venom and a short temper. In
this individual's defense, however, I must confess that I had been
poking it with a stick.

Surprisingly, the study of rainforest seeds can involve a lot of
snake-poking. There is a simple explanation for this: science loves a
straight line. Lines, and the relationships they imply, pop up everywhere, from chemistry to seismology, but for biologists the most
common line of all is the transect. Whether one is counting seeds,

FIGURE 1.1. A fer-de-lance *(Bothrops asper)*.
Anonymous (nineteenth-century). REPRODUCTION
© 1979 BY DOVER PUBLICATIONS.

surveying kangaroos, spotting butterflies, or searching for monkey
dung, following an arrow-straight transect across the landscape is
often the best way to make unbiased observations. They're great because they sample everything in their path, cutting directly through
swamps, thickets, thorn bushes, and anything else we might otherwise prefer to avoid. They're also horrible because they sample
everything in their path, cutting directly through swamps, thickets,
thorn bushes, and anything else we might otherwise prefer to avoid.
Including snakes.

Ahead of me, I heard the ring of machete on vine as my field assistant, José Masis, slashed us a path through the latest jungle imped-
iment. I had time to listen because the snake, having missed my boot
by inches, did something extremely disconcerting. It disappeared.
The mottled browns of a fer-de-lance's back make an excellent
camouflage, and I never would have seen so many of them—not to
mention eyelash vipers, hog-nosed pit vipers, and the occasional
boa constrictor—if I hadn't been diligently walking straight lines
through the forest, bent low to the ground, rummaging through the

leaf mulch. Some transects seemed to hold more snakes than seeds, and José and I developed techniques for nudging them out of the way or even lifting them on sticks and tossing them gently aside. Now, with an angry, invisible viper somewhere at my feet, new questions emerged. Was it best to stand still and hope the snake wasn't repositioning itself for another strike? Or should I run, and if so, in which direction? After a tense minute of indecision I ventured a step, then two. Soon I had resumed my seed transect without incident (though not before cutting myself a much longer snake-poking stick).

Scientific research often combines moments of excitement and discovery with long periods of monotonous repetition. More than an hour passed before my slow searching turned up the day's reward. There, directly in the path before me, sprouted a seedling of the great *almendro*, a towering tree whose fascinating natural history had

FIGURE 1.2. The sprouting seed of an *almendro* tree (*Dipteryx panamensis*). PHOTO © 2006 BY THOR HANSON.

drawn me to this rainforest in the first place. Though unrelated to
the nut trees of North America and Europe, the name translates
to "almond," a reference to the fatty seeds at the center of each fruit.
I noted the tiny plant's size and location in my field book, and then
crouched down for a closer look.

The seed's shell, so difficult to open in the lab, lay upended in
halves, neatly split by the pressure of the growing sprout. A dark
stem arched downward into the soil, and above it two seed leaves
had begun to unfurl. They looked impossibly green and tender, a
rich meal for the pale shoot just visible between them. Somehow,
this tiny speck had the potential to reach the forest canopy far above
me, its first steps fueled entirely by the energy of the seed. That same
story repeated itself everywhere I looked. Plants lay at the heart of
the rainforest's great diversity, and the vast majority of them had
started out exactly this same way, the gift of a seed.

For the *almendro*, the transformation from seed to tree seemed
particularly incredible. Mature individuals often exceed 150 feet (45
meters) in height, with buttressed trunks 10 feet (3 meters) across
at the base. They live for centuries. Their iron-hard wood is known
to dull or even break chainsaws, and when they flower, vivid purple
blossoms festoon their crowns and rain down to carpet the ground
below. (In my first scientific presentation on the tree, I lacked a de-
cent photo of its flowers, but got the point across with the closest
approximation of the color that I could find: a Marge Simpson wig.)
Almendro trees produce so much fruit they are considered a keystone
species, vital to the diets of everything from monkeys to squirrels to
the critically endangered Great Green Macaw. Their loss could alter
the ecology of a forest, leading to a cascade of changes, and even to
local extinctions for the species that depend upon them.

I was studying the *almendro* because throughout its range, from
Colombia north to Nicaragua, the tree has faced increasing chal-
lenges as forests have been cleared for ranching and agriculture, and
as demand has increased for its dense, high-quality timber. My research focused on the survival of *almendros* in Central America's

rapidly developing rural landscape. Could it persist in small fragments of rainforest? Would the flowers still be pollinated, the seeds be dispersed, and the next generations be genetically viable? Or were the majestic old trees that were now isolated in pastures and forest patches merely "the living dead"? If these giants couldn't reproduce successfully, then all of their complex relationships with other forest species would begin to unravel.

The answers to my questions lay in the seeds. So long as José and I could find enough of them, their genetics could tell us the rest of the story. Every seed and seedling we encountered held clues about its parents coded into its DNA. By carefully sampling and mapping them in relation to the adult *almendros*, I hoped to find out just which trees were breeding, where their seeds were going, and how those things changed as the forest was carved into fragments. The project lasted for years and involved six trips to the tropics, thousands of specimens, and countless hours in a laboratory. At the end I had a dissertation, several journal articles, and some surprisingly hopeful news about the future of the *almendro* tree. But only after all the samples were analyzed, the papers written, and the diploma delivered did I realize that something fundamental was missing. I still didn't really understand how seeds worked.

Years passed, other research projects came and went, but still this mystery puzzled me. While everyone from gardeners and farmers to the characters in children's books trust that seeds will grow, what makes it happen? What lies inside those neat packages just waiting for the spark to build a new plant? When I finally decided to get to the bottom of these questions, my mind immediately pictured that sprouting *almendro* tree, and how every part of its big seed was clearly visible, like the image in a textbook. Popping down to Costa Rica to find a fresh one was out of the question, but *almendro* is far from the only species with large, easily sprouted seeds. In fact, nearly any grocery store, fruit stand, or Mexican restaurant keeps the big seeds (and surrounding fruit) of at least one rainforest tree always in good supply.

FIGURE 1.3. Domesticated in Mexico and Central America over 9,000 years ago, avocados were long established in local diets by the time of the Aztec feast pictured here. Anonymous (Florentine Codex, late sixteenth century). WIKIMEDIA COMMONS.

In a brilliant piece of casting work, the title role for the movie "*Oh, God!*" went to George Burns. When asked about his greatest mistakes, the Burns-almighty deadpanned a quick response, "Avocados. I should have made the pits smaller." Sous-chefs in charge of guacamole would certainly agree, but to botany teachers around the world, the avocado pit is perfect. Inside its thin brown skin, all the elements of the seed are laid out in large format. Anyone wanting a front-row seat for a lesson on germination needs nothing more than a clean avocado pit, three toothpicks, and a glass of water. The simplicity of it was not lost on early farmers, who domesticated the avocado at least three different times from the rainforests of southern

Mexico and Guatemala. Long before the rise of the Aztecs or the Mayans, people in Central America already enjoyed a diet rich with the creamy flesh of avocado. I enjoyed it, too, binging on a spree of delicious sandwiches and nachos in preparation for my experiment. With a dozen fresh pits and a handful of toothpicks, I headed for the Raccoon Shack to get started.

The Raccoon Shack sits in our orchard, an old shed sided with tar paper and scrap lumber, and named for its former inhabitants. The raccoons once made an easy living there, gorging themselves on our apple harvest every fall. We had to give them notice, however, when parenthood suddenly required me to find an office space out-side the confines of our small home. The shack now boasts power, a woodstove, a hose spigot, and plenty of shelf space—everything I might need to coax my avocados to life. But I wanted more than germination; I knew to expect roots and greenery. What I needed was to understand just what inside that seed was making it all hap-pen, and how such an elaborate system evolved in the first place. Fortunately, I knew just the people to talk to.

Carol and Jerry Baskin met on the first day of graduate school at Vanderbilt University, where they both enrolled to study bot-any in the mid-1960s. "We started dating right away," Carol told me, so they were seated next to one another when the professor came around assigning research topics. In pairs. "That was special because it was the first time we worked together," she remembered. It also marked the first time they turned their minds to a topic that would define their careers. Though they insist that their romance was typical—mutual friends, similar interests—there has been noth-ing standard about the intellectual partnership it fostered. Carol fin-ished her doctorate a year ahead of Jerry, but they've been pretty much in synch ever since, publishing more than 450 scientific arti-cles, chapters, and books on seeds. For a guided tour of an avocado pit, no one on the planet could have better credentials.

"I tell my students that a seed is a baby plant, in a box, with its lunch," Carol said at the start of our conversation. She speaks with a southern drawl and has a casual way of explaining things, talking around the edges of difficult concepts until the answers seem to reveal themselves. It's easy to see why students rank her among the best science teachers at the University of Kentucky. I reached Carol by phone in her office, a windowless room where piles of papers and books cover every surface and overflow into the lab next door. (Jerry had recently retired from the same department, which apparently involved moving his piles of books and papers home to their kitchen table. "There are just two little clear places where we eat," Carol laughed. "It's a problem if we want to have company.")

With her "baby in a box" analogy, Carol neatly captured the essence of seeds: portable, protected, and well-nourished. "But because I'm a seed biologist," she went on, "I like to take things a step further: some of those babies have eaten all their lunch, some have eaten part of it, and some haven't even taken a bite." Now Carol opened a window onto the kinds of complexities that have kept her and Jerry fascinated for nearly five decades. "Your avocado pit," she added knowingly, "has eaten all of its lunch."

A seed contains three basic elements: the embryo of a plant (the baby), a seed coat (the box), and some kind of nutritive tissue (the lunch). Typically, the box opens up at germination, and the embryo feeds on the lunch while it sends down a root and sprouts up its first green leaves. But it's also common for the baby to eat its lunch ahead of time, transferring all of that energy to one or more incipient leaves called *seed leaves*, or *cotyledons*. These are the familiar halves of a peanut, walnut, or bean—embryonic leaves so large they take up most of the seed. As we were talking, I plucked an avocado pit from the pile on my desk and split it open with my thumbnail. Inside, I could see what she meant. The pale, nut-like cotyledons filled each half, surrounding a tiny nub that held the fledgling root and shoot. For a seed coat, the pit offered little more than varnish—thin, papery stuff that was already flaking off in brown sheets.

"Jerry and I study how seeds interact with their environment," Carol said. "Why seeds do what they do when they do it." She went

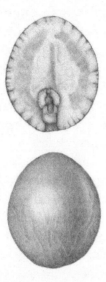

FIGURE 1.4. Avocado (*Persea americana*). Inside the paper-thin seed coat of an avocado pit, two massive seed leaves surround a tiny nub containing the root and shoot. Avocados evolved in a rainforest, where young trees need a large dose of seed energy to sprout and get established in deep shade. ILLUSTRATION © 2014 BY SUZANNE OLIVE.

on to explain that the avocado's strategy is somewhat unusual. Most seeds dry out as they ripen, using a thick, protective coat to keep moisture at bay. Without water, the embryo's growth slows to a near standstill, a state of arrested development that can persist for months, years, or even centuries until conditions are right for germination. "But not avocados," she warned. "If you let those pits dry out, they'll die." The way Carol said this reminded me that my avocado pits were living things. Like all seeds, they are live plants that have simply put development on pause, waiting until they land in just the right place, at just the right time, to send down roots and grow.

For an avocado tree, the right place is somewhere its seeds will never desiccate and the season is always right for sprouting. Its strategy relies on constant warmth and dampness, conditions you might find in a tropical rainforest—or suspended over a glass of water in the Raccoon Shack. With no need to survive long droughts or cold winters, avocado seeds take only the briefest pause before trying to grow again. "The avocado's dormancy may simply be the time necessary for the process of germination to take place," Carol explained, "which shouldn't be all that long."

I tried to keep that in mind during the slow weeks before my avocado pits showed any signs of life. They became my silent, unchanging companions: two rows of mute brown lumps lined up on a bookshelf below the window. Although I have an advanced degree in botany, I also have a long history of killing houseplants, and I began to fear for them. But like any good scientist, I took comfort in data, filling an elaborate spreadsheet with numbers and notes. Though nothing ever changed, there was a certain satisfaction in handling every seed, dutifully monitoring its weight and dimensions.

When it happened, I didn't believe it. After twenty-nine inert days, Pit Number Three gained weight. I recalibrated the scale, but there it was again, the most encouraging tenth of an ounce I've ever measured. "Most seeds take up water right before they germinate," Carol confirmed, a process cheerfully known as *imbibing*. Why it often takes so long is the subject of debate. In some cases, water may need to breach a thick seed coat or wash away chemical inhibitors. Or the reason may be more subtle—part of a seed's strategy to differentiate brief rain showers from the sustained dampness necessary for plant growth. Whatever the reason, I felt like pouring a libation for myself as, one after another, all my avocado pits began doing it. Outwardly they looked the same, but inside, something was definitely going on.

"We know a little bit about what's happening in there, but not everything," Carol admitted. When a seed imbibes, it sets off a complex chain of events that launches the plant from dormancy straight into the most explosive growth period of its life. Technically, *germination* refers only to that instant of awakening between water uptake and the first cell expansion, but most people use the term more broadly. To gardeners, agriculturalists, and even the authors of dictionaries, germination includes the establishment of a primary root and the first green, photosynthetic leaves. In that sense, the seed's work isn't done until all of its stored nutrition is used up—that is, transferred to an independent young plant capable of making its own food.

My avocados had a long way to go, but within days the pits began splitting apart, their brown halves tilted outward by the swelling roots within. From a tiny nub in the embryo, each primary root grew at an astonishing pace—a pale, seeking thing that plunged downward and tripled in size in a matter of hours. Long before I saw any hint of greenery, every pit boasted a healthy root stretching to the bottom of its water glass. This was no coincidence. While other germination details vary, the importance of water is constant, and young plants place top priority on tapping a steady source. In fact, seeds come prepackaged for root growth—they don't even need to make new cells to do it. That may sound hard to believe, but it's similar to what clowns do with balloons all the time.

Scrape the side of a fresh avocado root, and you'll get thin, curly strips like the radish shavings on a fancy salad. I placed one of these under my microscope and saw lines of root cells in sharp relief—long, narrow tubes that looked a lot like the balloons a clown might use to tie animal shapes. And just like a clown, the embryonic roots stuffed inside of seeds know that you don't show up to a party with your balloons already inflated. Even oversized clown pockets couldn't possibly hold enough. Empty balloons, on the other hand, take up no space at all and can simply be filled with air (or water) whenever and wherever the need arises.

The difference in size between empty and inflated balloons is actually quite astounding. A standard package of "Schylling Animal Refills" from my local toy store contains four greens, four reds, five whites, and assorted blues, pinks, and oranges, for a total of twenty-four balloons. Deflated, they fit easily into my cupped hand: a bright, rubbery bundle less than three inches (seven and one-half centimeters) across. Once I began blowing them up, I quickly appreciated why any good clown also travels with a helium tank or a portable air compressor. Lightheaded and wheezing, I tied the last balloon forty-five minutes later and sat surrounded by a riot of color. The balloons now formed a squeaking, unruly pile four feet (one and one-quarter meters) long, two feet (sixty centimeters) wide, and a foot (thirty centimeters) tall. Lined up end to end, they stretched from my desk out the door, across the orchard, through the gate, and onto the lane, for a total of ninety-four feet (twenty-nine meters). Their volume had risen by a factor of nearly 1,000, with the potential to form a skinny tube 375 times longer than the rubbery ball I'd started with—all from the addition of air. Give water to a seed, and its root cells will fill to do virtually the same thing, stretching longer and longer as they inflate. The process can last for hours or even days—a massive burst of growth before the cells at the tips even bother dividing to make new material.

Seeking out water is an understandable plant priority. Without it, growth stalls, photosynthesis sputters, and nutrients can't be liberated from the soil. But seeds can have subtler reasons to start growing this way, and no example makes that case better than coffee. As everyone with an early-rising toddler knows, coffee beans contain a potent and very welcome blast of caffeine. But while it may be stimulating to weary mammals, caffeine is also known for getting in the way of cell division. In fact, it stops the process cold, a tool so effective that researchers use caffeine to manipulate the growth of everything from spiderworts to hamsters. In a coffee bean, this trait does wonders to maintain dormancy, but it poses a distinct problem when it's finally time to germinate. The solution? Sprouting coffee seeds shunt their imbibed water to both root and shoot, swelling them rapidly to propel their growing tips safely away from the stifling effects of the caffeinated bean.

Avocado pits contain a few mild toxins to ward off pests, but nothing that slows things down once the game is afoot. I watched the roots grow and branch for days before the first greenery appeared, a tiny shoot emerging from the widening crack at the top of each pit. "It's accurate to call the next phase a massive transfer of energy from the cotyledons," Carol told me, explaining how what started out as the seed's "lunch" would now fuel a surge of upward growth. Within a few weeks, I found myself the caretaker not of seeds, but of saplings, young trees that bore little resemblance to the pits I'd nurtured for months. As a parent, I was reminded of the many transformations I'd already witnessed in young Noah's life, and something that Carol had mentioned suddenly struck me. Early in their careers, she and Jerry had decided they were just too busy to have children. In studying seeds, I now realized, they had nonetheless devoted themselves to the fickle lives of babies.

The Baskins' decades of work illustrate just how much there is to learn about what happens inside a germinating seed. Questions raised over 2,000 years ago by Theophrastus, "the father of botany," continue to challenge scientists. As Aristotle's student and successor, Theophrastus led exhaustive plant studies at the Lyceum, publishing books that remained definitive for centuries. Working on everything from chickpeas to frankincense, he described germination in great detail, wondering about seed longevity as well as differences "in the seeds themselves, in the ground, in the state of the atmosphere, and in the season at which each is sown." In the long years since, researchers have unraveled many of the processes guiding dormancy, awakening, and growth. It is well established that germinating seeds imbibe water and extend their roots and/or shoots through cell expansion. This stage is followed by rapid cell division fueled by the energy in their food reserves. But the exact cues that trigger and coordinate these events retain an aura of mystery.

Germination chemistry alone involves a huge variety of reactions as the dormant metabolism comes to life, producing all the hormones, enzymes, and other compounds necessary to transform stored food into plant material. For avocados, that stored food includes everything from starch and protein to fatty oils and pure sugar—a mixture so rich that nurseries don't even bother with fertilizers until well beyond the seedling stage. Transferring my young trees to potting soil, I noticed their cotyledons still clinging to the bases of the stems like pairs of upraised hands. Months or even years after rooting and leafing out, young avocado trees can still eke out a trickle of energy from the lunches their mothers packed. It's no coincidence that an avocado endows its offspring so generously. Like *almendros*, avocados evolved to sprout in the deep shade of a rainforest, where light is scarce and where massive food reserves can give the seedlings a distinct advantage. Their story (and their seeds) would be entirely different if they had hailed from deserts or high mountain meadows, places where every young plant has a quick path to full sun.

Seed strategies vary incredibly, their shapes and sizes adapted to every nuance of habitat on the planet. While this makes them a fascinating topic for a book, it can also make it hard to agree on just what part of a plant constitutes the seed. For purists, the seed includes only the seed coat and what lies within. Everything outside of that is fruit. In practice, however, seeds often co-opt fruit tissues for protection or other seed-like roles, and their structures become so fused that they're difficult or impossible to distinguish. Even professional botanists often fall back on a more intuitive definition: the hard bit encompassing the baby plant. Or, even more simply: what a farmer sows to raise a crop. This functional approach equates a pine nut with a watermelon pip or a kernel of corn, avoiding technical distractions about the role of every plant tissue involved. It's a model well suited to this book, but not without noting just how strangely different the contents of seeds can be.

Because the products of evolution work so beautifully in practice, it's easy to imagine the process chugging along like some grand assembly line, fitting each cog and sprocket to its particular place, for its particular function. But as any fan of *Junkyard Wars*, *MacGyver*, or Rube Goldberg devices knows, common objects can be reimagined and repurposed, and almost anything will work in a pinch. The sheer ceaselessness of natural selection's trial and error means that all sorts of adaptations are possible. A seed may be a baby in a box with its lunch, but plants have come up with countless ways to play out those roles. It's like a symphony orchestra. Violins get the melody most of the time, but there are also bassoons, oboes, chimes, and two dozen other instruments perfectly capable of carrying a tune. Mahler favored the French horn, Mozart often wrote for flutes, and in Beethoven's *Fifth Symphony*, even the kettledrums get a crack at that famous da-da-da-*dum*!

With their two hefty cotyledons, avocados illustrate a very common seed type, but grasses, lilies, and a number of other familiar plants have only one cotyledon, while pine trees boast up to twenty-four. In terms of lunch, most seeds use a nutritious product of pollination called *endosperm*, but various other tissues will do the job, including *perisperm* (yucca, coffee), *hypocotyl* (Brazil nut), or the *megagametophyte* preferred by conifers. Orchids don't pack a lunch at all—their seeds simply pilfer the food they need from fungi found in the soil. A seed coat can be papery thin, like an avocado's, or thick and hard, like those found inside pumpkins, squashes, and gourds. Mistletoes, in contrast, have replaced their seed coats with a mucilaginous goop, while many other seeds co-opt the hardened inner layers of the surrounding fruit. Even something so basic as the number of babies in the box can vary, with species from Lisbon lemons to prickly-pear cacti sometimes stuffing multiple embryos into a single seed.

Distinctions among seed types define many of the major divisions in the plant kingdom, and we'll touch on them again in later chapters as well as in the glossary and notes. Most of this book, however, focuses on traits that *unite* seeds, joining them in the common goals of protecting, dispersing, and feeding baby plants. Of these, nothing is more intuitive than the last, because, as everyone knows, the food in seeds gets eaten by a lot more things than baby plants.

In the Costa Rican forests where José and I worked, we often headed for the closest *almendro* tree to take our lunch break. Their huge, buttressed roots provided a good backrest, and their spreading canopies helped to shelter us from both sun and rain. But, just as importantly, *almendros* were the best places around to see wildlife. The stony shells of old seeds littered the ground beneath them in all states of disrepair, split apart by parrots feeding above or gnawed open by various large rodents. When peccaries approached, we always heard them coming, rattling whole seeds against their teeth as they positioned them for a cleaving bite. The sound was like billiard balls clacking against each other.

Raw *almendro* seeds always struck me as a bit mealy and bland. But when Eliza and I once roasted a panful, their sweet nutty scent filled the whole house, and the flavor wasn't half bad. With a little selective breeding to make the shells more cooperative, I could see them finding a place alongside the walnuts and filberts in our pantry. After all, that kind of experimentation is exactly the process that brought nuts, legumes, grains, and countless other seeds into human larders around the globe. When it comes to stealing the food from baby plants, no animal is more accomplished than *Homo sapiens*, and the importance of seeds in the human diet can't be overstated. We take them everywhere we go, planting them, nurturing them, and devoting whole landscapes to their production. As Carol Baskin put it, "when people ask me why seeds matter, I have one question for them: 'What did you eat for breakfast?'" Chances are, that meal began in a field of grass.

CHAPTER TWO

The Staff of Life

Behold, I have given you every plant yielding seed that is on the face of all the earth, and every tree with seed in its fruit: You shall have them for food.

—Genesis 1:29

South Dakota's Mount Rushmore boasts the huge granite heads of four US presidents. Hillsides in England sometimes feature prehistoric figures—great giants or running horses etched with trenches of chalk. China's Carved Hills of Dazu harbor thousands of ornate Buddhist sculptures, while the sprawling shapes scattered across Peru's Nazca Province include monkeys, spiders, a condor, and graceful spirals large enough to be seen from space. In Idaho, the hills have eyebrows. But while that may not sound as grand as giants or presidents, Idaho's eyebrows rank among the rarest landscape features anywhere.

Standing in the middle of one with my eyes closed, I held a plot frame out in front of me, turned a quick circle, and tossed it. The frame landed with a swish on the steep slope, a plastic rectangle that now enclosed one randomly selected square foot of an endangered ecosystem: Palouse Prairie. I knelt down beside it, opened my notebook, and began to count. The page soon filled as I checked off nearly twenty different plants crowded into that tiny patch of ground. I saw forget-me-nots, irises, paintbrushes, and asters, but above all there were the grasses—dense green tufts of fescue and delicate June grass waving in the breeze. You don't need a botany degree to know that native prairies are a good place to grow grass. That is their glory and their downfall, because nothing is more important to the human endeavor than the growing of grass.

Proof of that statement lay all around me, just beyond the edge of the eyebrow, where the jumble of prairie plants gave way to cultivated green fields that stretched to the horizon. They contained grass, too—a tall, Middle Eastern species in the genus *Triticum* that we know by the name of wheat. All over the world, wherever people go, they take wheat with them; it's an essential crop now grown on more land than all of France, Germany, Spain, Poland, Italy, and Greece combined. When European settlers reached the Palouse region of northern Idaho and adjacent Washington State, they immediately realized its potential. Formed from ancient windblown sediments, the Palouse's rolling, dune-like hills held a topsoil ideally suited to grains, a natural grassland where no irrigation was necessary. Plow met prairie quickly, transforming the area into a top wheat producer within a single generation. The few scraps of original prairie that remain lie in places too difficult to cultivate, strung out just below the rims of the steepest hillsides. From a distance, they look like thin, dark lines edging the curved tops of each hill, as if the landscape were raising its grassy "eyebrows" in surprise.

My plant surveys provided the botanical backdrop for an interdisciplinary team of entomologists, soil and worm specialists, and social scientists. The project aimed to better understand and protect the last Palouse prairies and to raise their profile in the local community—grass pride in a grass place. It gave me a crash course in identifying fescues, bromes, wild oats, cheat, and bluegrasses. Every hour in the eyebrows led to even more time at the microscope as I learned the different species by the subtleties in their leaves and by the various hairs, ridges, and wrinkles adorning their flower parts

FIGURE 2.1. The rolling, dune-like hills of the Palouse support fragments of native prairie amid one of the richest grain-producing landscapes in the world. WIKIMEDIA COMMONS.

and seeds. But while working in the Palouse taught me about grass diversity, it left an even stronger impression of how grasses, and particularly their seeds, have shaped human societies.

For tourists, grain elevators rising over a farm town represent one of the quintessential Palouse photo opportunities. To locals, they are the economy incarnate—brimming full with the seeds of a good crop, or looming as empty reminders in times of want. During the fall harvest, school attendance drops and the banks in town adjust their signs to alternate between the hour, the temperature, and the spot price for wheat futures. Versions of this same story play out in wheat country everywhere, from the plains of central China to the Argentinean pampas to the irrigated shores of the Middle Nile. And wheat does not stand alone as a grass crop of influence. Corn, oats, barley, rye, millet, and sorghum are also grasses, not to mention rice, the foundation of Asian diets for millennia. In Japan, Thailand, and parts of China, local words for rice can have telling double meanings: "meal," "hungry," or simply "food." Taken together, grains provide more than half of all calories in the human diet and take up more than 70 percent of the land in cultivation. They include three of the top five agricultural commodities, and they also bulk up the feeds used to fatten domestic cattle, poultry, swine, and even farmed prawns and salmon. When the prophet Ezekiel predicted famine in Jerusalem, he said that God would "break the staff of bread." By the seventeenth century, the phrase "staff of life" had come into use for all the staple grains, or the breads made from them. In the twenty-first century, little has changed: grass seeds still feed the world.

The strong ties binding people and grasses date to the roots of agriculture itself, when plant gatherers began to choose and manipulate their staples from the myriad wild species around them. Grains figured prominently in the founding of virtually every early civilization: barley, wheat, and rye in the Fertile Crescent (10,000 years ago), rice in China (8,000 years ago), corn in the Americas (5,000 to 8,000 years ago), and sorghum and millet in Africa (4,000 to 7,000 years ago). Some think the human reliance on grains (and other seeds) began much earlier, but regardless of when it started, our grass habit relies on specific traits found within the seeds. Unlike an avocado pit, where fat cotyledons fuel slow, steady growth in the shade, grass seeds evolved for life on the plains, where a quick start is the key to success. They are tiny and prolific and eager to sprout, traits that make grasses an ideal food crop and a dominant plant on virtually any patch of open ground. No toothpicks or water glasses are necessary to watch grass seeds grow. A woodpile and a January rainstorm will do the job nicely.

Everyone needs a hobby. Biologists, however, often run the risk of redundancy on their days off. Does it count as a vacation when I head outside to watch birds, catch bees, or look at plants? I do play bass in a jazz band, but anyone tallying my free moments would find one thing taking up more time than any other: firewood. We live in a 1910 farmhouse that was once sawn in half, loaded onto a flatbed truck, and hauled five miles down a country road to its present location. Patching it back together resulted in a charming but drafty structure that no amount of fiberglass batting can properly insulate. As a result, it's a rare day that doesn't find me sawing, splitting, stacking, or restacking some portion of the four cords we burn every year to cook and keep warm.

Finding all that fuel has turned me into a bona fide wood scrounge, combing the roadsides after every windstorm, and pestering neighbors and relations for tips on surplus timber. I'll take anything, and so I was happy to help clean up the old madrona logs cluttering a friend's yard. Madrona trees belong in the heather family and look like giant rhododendrons, their curving trunks and branches covered in a beautiful reddish bark. As I set to work, it struck me as odd that this madrona was decidedly green. Looking closely, I soon saw why. Having sat outside for more than a year, surrounded by tall meadow grasses, these logs and branches had collected seeds in every crack and crevice. And now they were beginning to sprout. Swollen with recent rains, each tiny grain had sent up a spear of purest green, giving the surface of the wood a decidedly fuzzy, grassy look. If the Chia Pet folks offered their wares in a woodpile motif, this would have been it.

I pulled up one of the little grasses and found the faintest husk of a seed remaining, a thin, split thing where the base of the green shoot turned pale. Rather than invest in fat cotyledons, grasses endow their offspring with only a modest lunch and rely instead on fecundity—broadcasting droves of seed in the hope that a few will find purchase. Where a pampered avocado tree might yield 150 single-seeded fruits annually, I recently counted 965 seeds on the wispiest looking bent-grass growing in our driveway. The food stored in a grass seed gives that baby plant enough energy for a quick growth spurt, but would never keep it alive for long in the shade. Instead, young grasses depend on finding open, unoccupied real estate. They prefer soil, but will also germinate on pavement, in gutters, or on the running boards of old pickup trucks. Some species thrive in sand or on mud flats, or make a quick living colonizing the shifting gravel of riverbanks. To rock climbers everywhere, grass seeds create the constant need for "gardening," yanking tufts of new greenery from the tiny cracks and crevices that both climbers and plants hope to cling to.

Contrary to popular belief, watching grass grow can actually be quite riveting—a story of derring-do combined with sheer tenacity. Though I hated passing up free firewood, I left a stack of the grassy madrona in place to let the drama unfold. Six months later, the meadow lay baked in summer sunshine. I returned to find the logs in their pile, but with hardly a trace of that promising green fringe. Nearly every seedling had withered in the heat, exhausting its tiny lunch long before a root could stretch down to reliable water. But one plant had survived. From the split end of a log near the bottom of the pile, a tuft of velvet grass now sprouted, its tall flower stalk rising upward to sway in the breeze. I carefully lifted the wood and saw where roots had threaded their way through a crevice to find the soil below. In general, scattering seeds on woodpiles spells a death sentence for the baby plants inside. But this one success story, with the hundreds of seeds it would produce, helped justify the tactic.

While a grass's profligacy might not have the cozy, nurturing appeal of the well-stocked lunches found in avocados, nuts, legumes, and other plump seeds, it's certainly a successful strategy. Programmed for colonization and survival, the tiny grains of most species can withstand desiccation and long periods of dormancy, traits that help grasses dominate almost every earthly habitat too arid for trees and shrubs. Even Antarctica boasts native grasses, and if you lined up all the flowering plants on earth, nearly one in twenty would be a grass. Ubiquity alone, however, does not a staple make. For all their prolific seed, grasses would hardly be so vital to people without a trick of chemistry. That trick lies in the way they pack their lunch.

Dissecting a grass seed takes steady hands, and if you decide to try it, I suggest skipping the afternoon coffee. My jittery attempts sent half a dozen wheat grains skittering off the desk before a clean slice finally revealed the pertinent feature: a mass of starch granules that glowed like lumpy marbles in the light of my microscope. Of all the ways that seeds can store energy, from oils and fats to proteins, none makes a better staple food for people than starch. It's built from long chains of glucose molecules, like sugar beads on a flimsy necklace. Enzymes in the human intestine, and even in our saliva, can break that necklace easily and release the sugars. Tweak the chemistry of starch ever so slightly, however, and you get cellulose, the indigestible plant fiber that makes up stems, twigs, and the trunks of trees. Cellulose and starch differ only in how their glucose chains hold together, a few repositioned atoms that change the flimsy string to an indigestible one, like a steel wire. Without starch's weak glucose bonds, and our ability to break them, grass seeds would pass through the human gut like a handful of sawdust. As it stands, the starch content of a grass seed can top 70 percent, quick energy that evolved to fuel plant growth but that now fuels over half of all human activity.

Given the abundance of grasses and their prolific, starchy seeds, it's not surprising that our ancestors learned to take advantage of them. It seems that wherever people made the switch from hunting and gathering to cultivation, a grass or two lay at the heart of it. The civilizations that followed helped to cement our dependence on grass calories, and those few select species then spread to fields and garden plots around the world. But while historians have long considered the grain diet a relatively recent phenomenon, a product of the agricultural revolution, new thinking puts the seeds of grasses and other plants high on the human menu far into our nomadic, hunter-gatherer past.

"It's absolutely reasonable to assume that seeds have been part of the diet forever," Richard Wrangham told me. "After all, chimpanzees eat them." As a professor of biological anthropology at Harvard University, Wrangham should know: he published his first paper on chimp feeding habits in the early 1970s, and has been studying them in the wild ever since. I first met Wrangham at a primatology workshop in Uganda, where he and Jane Goodall kicked off the keynote address with a piercing exchange of chimpanzee pant-hoots and shrieks. Two decades later, he still knows how to get people's attention. I called him at his Harvard office, where in spite of pressing research deadlines and a full teaching load, he sounded eager to explain his unorthodox new theory.

"I used to try and eat what the chimps ate," he began, recalling his early fieldwork in Uganda's Kibale Forest, "and I can tell you I was pretty hungry by the end of the day." At first, Wrangham assumed that he just wasn't suited to the fruits, nuts, leaves, seeds, and occasional raw monkey that made up the chimpanzee diet. But when he put his observations in the context of human evolution, a profound new idea emerged. It wasn't the type of food that mattered, but how it was prepared. "I became convinced that we cannot survive in the wild on raw food. As a species, we are entirely dependent on using fire for food preparation. We are the cooking ape."

In spite of his bold ideas, Wrangham speaks carefully, building a case with the patience of someone accustomed to long hours of observation in the field. "My perspective comes from working with apes. I see humans as apes that have been modified," he explained, noting our substantially smaller teeth, shorter intestines, and larger brains. He told me about the remarkable energy gain achieved through cooking—how roasting or boiling meats, nuts, tubers, and other primate foods increased digestibility by anywhere from a third for wheat and oats to as much as 78 percent for a chicken egg. Wrangham's theory proposes cooking as the critical innovation separating advanced members of the genus *Homo* from their more ape-like ancestors. By shifting to a highly digestible, cooked diet, our forebears no longer needed the massive molars and expansive guts that apes need to process fibrous raw foods. And with so much more energy available, we could suddenly afford the metabolic demands of a larger brain.

Though still controversial, the logic of Wrangham's thinking rings like a bell tone through the din of competing hypotheses. Traditionally, most anthropologists have emphasized the spear and arrow side of the hunter-gatherer equation, attributing changes in dentition and brain size to better hunting techniques and a protein-rich diet. But Wrangham maintains that no quantity of raw meat (or other raw foods) could adequately nourish modern hominids, let alone spur their evolution. "On a strictly raw diet," he explained, "you don't have time for high-risk activities like hunting. If our ancestors ate like chimpanzees, they would have had to spend at least six hours a day just sitting around and chewing."

By deemphasizing the importance of meat, the cooking-ape theory raises the profile of the gatherers, who brought in a much wider range of foods, from roots and honey to fruit, nuts, and seeds. "Tubers were probably the fallback," Wrangham mused, "but any rich seed would have been preferred when they could get it." He noted how chimps seek out the roasted beans of *Afzelia* trees following forest fires, as well as the common indigenous practice of abandoning hunting during seasonal bonanzas of prime fruit, nuts, or honey. Just when grains became a staple, however, remains uncertain. Their caloric potential is huge, particularly when they are cooked, but they require an organized harvest and considerable processing. A definitive answer requires more evidence, or, as Wrangham puts it, "encouragement from archaeology." Now that people are looking, however, that encouragement seems to be turning up everywhere.

For anyone interested in early human societies, the habits of recent hunter-gatherers provide an invaluable comparison. Groups from warm climates traditionally derived between 40 and 60 percent of their calories from plant foods. Many relied on grass seeds, and not just those from the wild ancestors of familiar crops like wheat or rice. In Australia, Aboriginal societies made bread and porridge from grasses as diverse as arm millet, hairy panic, mulga grass, star grass, ray grass, and naked woollybutt. Native Americans near what is now Los Angeles harvested canary grass right up to the Spanish missionary period, and maygrass, a close relative, provided starch to tribal diets throughout the eastern seaboard. People near the Sea of Galilee were using stone tools to grind and process wild barley over

20,000 years ago, and in Mozambique, similar methods put sorghum on the menu 105,000 years ago. But perhaps the most tantalizing ancient grains of all come from Gesher Benot Ya'aqov in Israel, where signs of controlled fire date back 790,000 years. There, nestled amid the scrapers and charred flint, researchers unearthed a tiny handful of burnt seeds: feather grass, goat grass, wild oat, and barley. That discovery puts edible grains by the fireside deep in the era of *Homo erectus*, hundreds of millennia before our species even evolved.

If Richard Wrangham is right, and this kind of archaeological encouragement continues, we may find that the caloric boost from eating cooked grains played a significant role in human evolution. But regardless of when grass seeds entered the diet, they were firmly established as staples by the time we settled down to farm. When that happened, however, our ancestors did away with the mulgas and woollybutts to concentrate on a few promising varieties. No place on earth illustrates that transition better than the ancient settlement of Tell Abu Hureyra, near the modern city of Aleppo, Syria. What started as a seasonal village for hunter-gatherers grew over time to an agricultural town with 4,000 to 6,000 permanent residents. Each era left behind a clear record of its activities, beautifully preserved in layers of sediment and debris. The early inhabitants enjoyed a diet of more than 250 different wild plant foods, with 120 kinds of seeds, including at least 34 different grasses. By the time farming was firmly established, however, that panoply had shrunk to lentils, chickpeas, and a few varieties of wheat, rye, and barley.

This same pattern repeated itself whenever and wherever the agricultural revolution took hold: diverse wild diets gave way to ones focused on a few staple grains and other crops. With few exceptions, the chosen grasses share several important traits. They are annuals, an all-or-nothing life strategy that encourages plants to put their resources into seed production. With only one growing season in which to live and reproduce, annuals have no need to invest in lasting stems and leaves but every reason to produce the kind of large, prolific seeds that make them an appealing target for cultivation.

In fact, the simple availability of large-seeded annuals may be the best predictor for the advent of agriculture. Thirty-two of the world's fifty-six heaviest-seeded grasses occur in the Fertile Crescent or other parts of Eurasia's Mediterranean zone, where many of the earliest civilizations flourished. As geographer Jared Diamond has observed, "that fact alone goes a long way toward explaining the course of human history."

Diamond has argued that the presence of readily domesticated grasses gave the Mediterranean region an environmental edge, helping its people develop early and dominant civilizations. The relative scarcity of such grains in parts of Africa, Australia, and the Americas may have delayed their journey toward agriculture, a hindrance with significant consequences for their later interactions with European and Asian cultures. Regardless of when the transition took place, however, those fundamental links between grass and civilization have never gone away. Once established as the basis of our diet, grains became thoroughly enmeshed in economies, traditions, politics, and daily life around the world. Any examination of history quickly finds grain at the root of transformational events.

In the latter days of the Roman Republic, city leaders pacified a restive public through lavish entertainments and free or heavily subsidized allotments of wheat, a strategy of diversions that Juvenal famously dubbed "bread and circuses." First codified by the Grain Law of Gaius Gracchus, grain subsidies continued for centuries as an important political tool throughout the empire. A goddess, Annona, was specifically invented by the state to personify the grain dole. She often appeared in statuary and on coinage, holding sheaves of wheat and perched on the prow of a ship to symbolize the steady arrival of grain into the capital. While historians attribute Rome's ultimate demise to everything from inflation to the mental-health drawbacks of lead plumbing, no one disputes that grain shortages hastened its fall. Long dependent on imports from North Africa, Rome first saw Egyptian production diverted to Constantinople, and then lost the rest of its supplies when Carthage fell to the Vandals.

FIGURE 2.2. Invented to embody the annual gift of grain from government to citizens, the goddess Annona was an early example of Roman propaganda. These coins, from the third century AD, show her holding grain sheaves and a cornucopia. On the left, her foot rests on the curved prow of an incoming grain ship; on the right she stands beside an overflowing *modius*, the basket used to dole out individual shares. PHOTO © 2014 BY ILYA ZLOBIN.

Prices skyrocketed, and major food riots and famines rocked the capital at least fourteen times in the fourth and fifth centuries. When the Visigoths laid siege to Rome in AD 408, the grain dole was cut in half and then to a third before the city was finally overrun. As two noted historians put it, "bread, or the lack of it, had finally destroyed the Western Empire."

When the Black Death raged through Asia and Europe in the fourteenth century, a baffled and terrified populace blamed it on everything from earthquakes to acne. Only later did epidemiologists trace the disease to tiny fleas inhabiting the fur of the common black rat. But even this insight failed to explain the plague's spread. After all, the average rat travels only a few hundred yards from its birthplace over the course of a lifetime, so how did the illness move from China to India and the Middle East and all the way north to Scandinavia in a matter of years? The answer lies not in the ranging habits of rats, but in their diet. While black rats will eat almost anything, they thrive on grain of all types and travel with it wherever it goes. And while most fleas live only a matter of weeks, those found in rat fur can persist for a year or more, and their larvae have learned to eat grain. So even on a long ship voyage, when all the plague-sick rats might die at sea, the fleas survived (with their offspring happily munching away in the hold), ready to infect new rats and people at every port of call. And though a fastidious overland merchant might rid his caravan of rats, there again were the fleas, safely tucked away in every bushel of grain. At its peak, the plague's rapid spread suggests that it must have become airborne—transmitted directly between people through coughing and sneezing. But historians still believe it got its start with the grain trade, skipping only the remotest backwaters, or kingdoms like Poland that kept their borders firmly closed. Periodic outbreaks continued until well into the twentieth century, striking places like Glasgow, Liverpool, Sidney, and Bombay—all of them busy ports with an active commerce in grain.

The history of revolts and uprisings also turns on grass, with grain shortages often providing the spark that transforms resentments into open rebellion. When fourth-century Chinese emperor Hui of Jin was told that his subjects were starving for lack of rice, he reportedly asked, "Then why don't they eat meat?" Subsequently, Hui lost half his kingdom in the Wu Hu Uprising. And though historians doubt that Marie Antoinette ever said, "Let them eat cake," no one questions that wheat and bread shortages helped spark the French Revolution, the Russian Revolution, and the Spring of Nations in 1848, a conflagration that affected fifty countries in Europe and Latin America. The trend continues to the present day. It is no coincidence that the Arab Spring began in Tunisia, the world's largest per capita wheat consumer, in a year following heat waves, floods, fires, and crop failures in some of the world's top wheat-producing nations. Tunisian wheat imports dropped by nearly a fifth in 2011, prices spiked, and widespread food riots swept the country in the months leading up to the revolution. Protests and riots over grain prices also preceded the rebellions in Libya, Yemen, Syria, and Egypt, a country where the word *aish* means both bread and life. The Algerian government, in contrast, responded to the food crisis with a massive investment in grain, *increasing* wheat imports by over 40 percent in 2011, stabilizing prices, and constructing huge storage facilities to stockpile against future shortfalls. Though unrest continues to sweep the region, and bread riots rocked the new Egyptian government in 2012, the Algerian regime still stands.

Of course, no single factor led to the Arab Spring, but the underlying role of wheat prices brings grain politics full circle. More than 10,000 years after the hunter-gatherers at Abu Hureyra first turned to farming, the grasses they helped domesticate continue to shape history. The legacy of the gatherers is profound. In the Fertile Crescent and around the world, access to grain plays a subtle but pervasive role in the fate of nations: when harvests are poor, governments falter. (The same cannot be said about the legacy of the hunters. No empire has ever collapsed from a shortage of antelope.) But it doesn't take a revolution or a plague to see the influence of grass seeds on modern life. Nothing reveals their role in our culture more vividly than a visit to grain country at harvest time.

"You're looking at 2 million bushels of soft wheat," said Sam White. We were perched on the back of his pickup truck, peering in through the door of a cavernous building heaped to the rafters. Cool, dry air drifted over the sea of grain and brushed past our faces. I did some quick math. With prices hovering around nine dollars a bushel, the wholesale value of this one storage depot exceeded $18 million. Processed into flour and packed into five-pound bags, it would bring over $100 million at a grocery store. Bake that flour into bread, pretzels, Pop Tarts, Oreos, or any of the thousands of other wheat-based products, and the bill at checkout would rise still higher. Before Sam rolled the big door closed, I raised up my camera and snapped a picture, but it didn't turn out. The grain just looked like a pile of sand; there was nothing to show that it stood three stories tall and stretched the length of two football fields. Nor could a photograph explain that hundreds of similar sheds and silos dotted the landscape in all directions, every one of them filled to the brim.

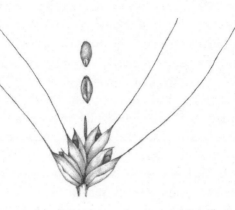

FIGURE 2.3. Wheat (*Triticum* spp.). Descended from wild grasses native to the Middle East, wheat now covers more agricultural acreage worldwide than any other crop. Like the individual grains of other edible grasses — from rice and corn to oats, millet, and sorghum — each tiny grain of wheat is actually a complete seed-like fruit called a *caryopsis*. ILLUSTRATION © 2014 BY SUZANNE OLIVE.

Anyone biting into a crusty baguette or twirling spaghetti noodles onto a fork has some vague notion that their meal began life on a farm. But few of us stop to consider the daunting logistics that lie between field and marketplace. The grain in Sam's barn was valuable, but the padlock on the door seemed superfluous—after all, who could possibly steal something that weighed 60,000 tons? Storing, processing, and moving grain on that scale requires infrastructure, and it was exactly that system of silos, trucks, roads, railroads, barges, and oceangoing freighters that I had returned to the Palouse to learn about.

"All the wheat is in now," Sam told me as we climbed back into his truck. "Barley too." My guide for the day, Sam White worked in upper management at the Pacific Northwest Farmers Cooperative, a group of 800 growers who co-own twenty-six storage and processing facilities in and around the small town of Genesee, Idaho (population 955). He grew up farming, but switched to the business side of things after college, and has now spent more than two decades selling Palouse grain in a complex global marketplace. Stocky, with sandy hair and a sun-weathered face, Sam likes his job: helping local farmers get the best prices for their crops. Not that it's always easy. "In my dad's time, if the price per bushel changed two cents in a year, that would be a big deal. Now you can see it swing thirty or forty cents in a day." On top of that, farmers form strong bonds with the crops they've worked so hard to nurture, and often let emotions cloud their business sense. "Frankly," he confided, "it's usually better if their wives decide when to sell."

We drove out of town between rolling fields of bright blond stubble lined with the tracks of combines. I smiled as we passed familiar prairie eyebrows, their coarse grasses and shrubs framing the hilltops like bushy parentheses. There was a wildfire burning in the nearby Bitterroot Mountains, and its smoke mingled with the dust billowing up behind distant tractors. It hadn't rained for months, but the fall planting was already well underway. Here and there, newly plowed fields stood out—broad swathes of dark earth awaiting their allotment of seed. Sam told me about tilling methods, fertilizers, and crop rotations, and then turned back to commerce: "Over 90 percent of what we grow here ends up in Asia," he said. That statistic surprised me, but it made perfect sense. Because in spite of its location, totally landlocked and more than 350 miles from the coast, the Palouse lies only a few minutes from a seaport.

Sam steered the truck onto a busy highway and we soon began heading sharply downhill. Every year, millions of tons of Palouse grain follow the same route, descending into a steep-walled canyon where the Clearwater River joins the Snake River at the town of Lewiston. There, we stood beneath towering concrete silos and a huge conveyor belt that cantilevered out over the river. It clanked and whirred above us, dropping a steady flow of wheat into the hold of a waiting barge. Golden chaff drifted through the air all around, glinting in the sunlight and settling like flotsam on the calm water.

"It takes three or four barges to make a tow!" Sam shouted above the noise, and went on to explain how the grain would make its way down the Snake and Columbia rivers all the way to the Pacific Ocean. Lewis and Clark traveled this same path in the early nineteenth century, but where the famed explorers navigated swift currents and treacherous rapids, modern vessels move through locks and dams and what is essentially a series of long, linear lakes. When journalist Blaine Harden boarded a tug to make the journey in the mid-1990s, his captain offered a sober prediction: "By the time you get to Portland, you are going to be bored shitless."

The Snake River dams, and the placid lakes behind them, may take the thrill out of river travel. But they also say something important about the political power of grain. Because while the dams on the Columbia supply massive irrigation schemes and produce half the region's electricity, water and hydropower were afterthoughts on the Snake. The four dams downstream of Lewiston were built to move cargo, and the cargo moving from Lewiston was grain. In 1945, awash in war debt, the United States Congress still deemed the transportation of Palouse wheat and barley a top government priority. It approved construction of "such dams as are necessary" to open up the lower Snake River to navigation, a massive infrastructure project that would last three decades and cost more than $4 billion in today's currency. At a ribbon-cutting ceremony in 1975, Idaho governor Cecil Andrus stood on the docks at Lewiston and predicted that his state's new seaport would "enrich our daily lives through international trade." The export boom that followed proved him right, and also proved that wheat and barley can do more than get dams built: they can also protect them in a changing political climate.

Within a few weeks of the governor's speech, rare fishes like Tennessee's infamous snail darter gained protection from the recently passed Endangered Species Act, and began complicating dam construction across the country. This trend reached Idaho in the 1990s, when four varieties of Snake River salmon and steelhead, decimated by the dams and slackwater, were added to the endangered species list. The resulting "Salmon Wars" illustrate how grain continues to hold sway in national politics. Breaching the Snake River dams became a rallying cry for fishing and environmental groups, and for a time seemed a likely outcome in efforts to save wild salmon. But though the idea of dam removal was bolstered by favorable court decisions and support from the likes of Vice President Al Gore, it slowly faded from the discussion. Instead, the government spent additional billions of dollars building fish ladders and hatcheries and even physically moving fish around the dams. Sometimes the little salmon go by tanker truck, but more often they travel the same way that grain does: by barge.

A few "SAVE OUR DAMS" signs can still be seen in Lewiston and nearby communities, but the lettering is faded and they seem redundant. Hardly anyone on either side now considers dam removal likely. When I asked Sam about the controversy, he said simply, "Dams are still an important part of how we move product." Sam struck me as a modest man, but that might have been his biggest understatement of the day: since Governor Andrus cut that ribbon in 1975, the Snake/Columbia system has become the third-busiest grain corridor in the world.

While multibillion-dollar dam schemes may sound extreme, they're far from the only way that politicians support growing, shipping, and marketing the grass seeds we all depend on. The same economic and cultural forces that led the Romans to invent Annona, the "free wheat" goddess, still keep governments around the world in the grain business. State-supported enterprises from Russia and Ukraine to Australia and Argentina continue investing heavily in transportation, export terminals, and subsidized production. In China, the practice dates back to at least the fifth century BC, when work began on the Grand Canal, a 1,104-mile-long (1,777-kilometer) waterway designed to keep wheat and rice supplies flowing to the capital. It's a timeless imperative. As agricultural lobbyists love to point out, skimping on a highway bill means a few more potholes; cutting the farm bill means that people don't eat.

Toward the end of our day on the Palouse, Sam and I passed a row of grain elevators on the edge of Genesee and came to a metal-sided building that hummed with activity. "Do you want to see some garbs?" he asked.

"Absolutely," I answered, hoping that I'd correctly understood "garbs" as an insider's slang for garbanzo beans.

Sure enough, I soon found myself on a noisy factory floor, dodging forklifts loaded high with legumes. We watched the garbs rattle down conveyor belts, pass through cleaners and sorters, and drop finally through an electronic eye that spotted any blemish and removed the offending bean with a blast of compressed air. Packed into hundred-pound sacks emblazoned with a "Clipper Brand" sailing ship, the final product was loaded into waiting trucks. Then it would either head west, to Seattle and the Asian ports beyond, or east, to a hummus factory in Virginia.

"Legumes are an important part of the rotation," Sam explained, after we'd left the clamor of the packing plant behind. "Most growers will do a fall wheat and a spring wheat, and then work in a crop of lentils, garbs, or split peas." Alternating crops helps keep the pest load down, but, just as importantly, the peas and beans fix nitrogen in the soil, naturally fertilizing the next grain crop. This pairing of grasses and legumes is as old as agriculture itself, a method repeated virtually everywhere that plant domestication has taken place. Garbanzos (or chickpeas), lentils, and split peas all developed alongside wheat and barley in the Fertile Crescent. In China, early rice farmers soon added soybeans, adzukis, and mung beans to the mix. Central America had its corn and pinto beans, while African millet and sorghum went hand in hand with cowpeas and groundnuts. More than just a good cropping method, this synergy extends all the way to the dining-room table, where starchy grains and protein-rich legumes complement one another perfectly, both in flavor and nutrition. The "complete proteins" found in combinations like rice and beans, or a lentil and barley salad, are common knowledge to anyone who has read the first page of a vegetarian cookbook. Essential nutrients that might be lacking in a particular grain can generally be found in the accompanying legume, and vice versa. But these stark differences in the contents of grains and legumes also raise very basic questions about the biology of seeds.

With grasses so successful in nature and so useful to people, it's obvious that packing one's seeds with a starchy lunch is a good evolutionary idea. So why don't all plants do it? Why do beans and nuts store energy in proteins and oils? Why does a palm kernel contain over 50 percent saturated fat? Why are jojoba seeds practically dripping with liquid wax? Grass starch may be the staff of life, but plants obviously have a lot of other ways to fuel their seeds and, by extension, us. Happily, one of the best ways to explore the range of sustenance packed into seeds involves a trip to the nearest candy aisle.

CHAPTER THREE
꙰

Sometimes You Feel
Like a Nut

God gives the nuts, but he does not crack them.

—German Proverb

In the late 1970s, the Peter Paul Manufacturing Company raised its suggested retail price for Almond Joy candy bars to twenty-five cents. But though this figure equaled my entire weekly allowance, I never regretted investing those wages in a confection the ad jingle summarized as "rich milk chocolate, coconut, and munchy nuts too!" At the time, it never occurred to me that my future career would reach this enviable moment: the opportunity to buy my favorite candy bars as a business expense. But a fact that escaped me then is extremely relevant now: from the first crunch of the roasted almond to the chewy sweetness of the chocolate and coconut finish, savoring an Almond Joy bar is an entirely seed-based experience. And while it's tempting to chalk up Almond Joys to the same logic that Benjamin Franklin used for beer—"proof that God loves us"— there's far more to their story. The seeds involved don't just taste good; they demonstrate beautifully the incredible range of ways that a plant can pack lunch for its offspring.

An Almond Joy now costs eighty-five cents at our local drug-store, and I've paid more than a dollar for them at vending machines. But you still feel like you're getting your money's worth because each package actually contains *two* small bars. This gives buyers the opportunity to share with a friend or save a piece for later, though it's unclear if anyone has ever done so. In my case, having two bars allowed me to eat one immediately and still have something left over to dissect. Cutting the bar in cross-section revealed its center of shredded coconut (from a pan-tropical palm), topped with almond (from an Asian tree in the rose family), and surrounded by a thin layer of chocolate (from a small New World rainforest tree). I took scrapes from each layer and prepared a microscope slide, but glancing at the package told me that none of these was the most dominant seed product in the bar. That honor rested with corn syrup, a sweetener derived from the seeds of a grass, maize, that is often used as a replacement for cane sugar (which, incidentally, also comes from a grass). But we already know from the last chapter that grasses are ubiquitous, and that their starch-filled seeds are easily transformed into sugars. The rest of the bar's contents tell us why seeds have developed so many other ways to store energy, and why we should all be thankful that they have.

The milk chocolate coating contained cocoa butter as well as a dark, bitter slurry that candy makers refer to as cocoa liquor, cocoa mass, or simply chocolate. These products both come directly from the large cotyledons found in a mature cacao bean. Squeeze the bean in a hot press and more than half its mass drips out as cocoa butter, a fat with the important quality of being solid at room temperature but liquid above approximately 90 degrees Fahrenheit (32 degrees Celsius). Since the average body temperature clocks in at 98.6°F, chocolate, quite literally, melts in your mouth. Roasting and milling the beans produces cocoa liquor, which can be mixed with varying amounts of cocoa butter, milk, and accompanying sweeteners to give us the wide range of chocolate flavors available in any well-stocked candy aisle. Farther down on the ingredients list, I spotted cocoa powder, another familiar cacao product, which comes from grinding the cake of dry "nibs" left over after pressing the beans for butter.

In the wild, cacao beans reside inside the fleshy pods of a small, shade-loving tree native to forests of southern Mexico, Central America, and the Amazon. I often stumbled upon old cacao orchards in Costa Rica while searching for *almendro* seeds. I would glance up from a transect to find myself suddenly surrounded by their pods—bizarre, gourd-like fruits that sprouted directly from trunks and branches in varying shades of orange, purple, chartreuse, and hot pink. It's no wonder that cacao caught the attention of the Mayans, Aztecs, and other early Americans, who developed a stimulating energy drink from the beans, and whose reverence for the species lives on in its genus name, *Theobroma*, "food of the gods." It took Europeans and the rest of the world a few centuries to really acquire the taste, but cacao trees now grow everywhere from Guatemala to Ghana, Togo, Malaysia, and Fiji, and global chocolate sales exceed $100 billion annually. The average German consumes more than twenty pounds of the stuff every year, and in Britain people spend more money on candy than they do on bread and tea. Ecologically, the extravagance of a large, rich bean makes perfect sense. Like *almendro* or avocado, cacao seeds evolved to sprout and grow in a dark forest, where young seedlings need large energy reserves to survive. But nothing I saw in cacao plantations, botany textbooks, or candy bars explained why that energy had to come in the form of fat instead of starch.

I turned to the next ingredient on the Almond Joy list, coconut, from a seed that ranks among the world's largest. Though familiar to anyone who has dreamed of palm trees and tropical beaches, the coconut is actually something of a mystery. Botanists call it *cosmopolitan*, a word that only came into common use in the nineteenth century, when global empires and fast sailing ships made it suddenly possible for an individual to become familiar with all parts of the world. For a plant, there can hardly be a greater compliment: so widespread and successful that no one is even sure where you came from.

FIGURE 3.1. Coconut (*Cocos nucifera*). The seeds of the coconut palm, among the world's largest, provide everything from thirst-quenching beverages to cooking oil, skin creams, and mosquito repellant. Dispersed throughout the coastal tropics by ocean currents and people, the origin of the species remains mysterious. ILLUSTRATION © 2014 BY SUZANNE OLIVE.

The coconut palm achieved this feat with fruits that function as massive, floating seeds. Each buoyant husk surrounds a single fist-sized kernel that is hollow except for a nutritious liquid known to health-food enthusiasts as "coconut water." Whatever branding specialist coined that term cannot be blamed for shying away from the more accurate, technical description: *acellular endosperm*. But while "endosperm" might not sound catchy in an ad campaign, its market potential should not be underestimated. As a coconut seed matures, much of its liquid hardens into a solid endosperm called *copra*, the familiar white flesh that graces not only candy bars and cream pies but also Filipino stews, Jamaican breads, and South Indian chutneys. Squeeze water through that flesh, and you get coconut milk, an essential ingredient in curries and sauces throughout the coastal tropics. And with minimal processing, copra yields over half its volume in coconut oil, one of the top five vegetable fats in the world and a common additive in everything from margarine to sunscreen.

To a Hollywood set designer, coconuts provide a reliable fallback prop for any tropical situation. They've been featured as drinking cups in productions ranging from *The Brady Bunch* to *Lord of the Flies*, and as bra cups in *King Kong*, *South Pacific*, and the Elvis blockbuster *Blue Hawaii*. The Professor, a character in the 1960s sitcom *Gilligan's Island*, famously used coconuts to build useful items like battery chargers and a lie detector. His inventions hardly seem exaggerated in light of the actual products made from coconuts, which include buttons, soap, charcoal, potting soil, rope, fabric, fishing line, floor mats, musical instruments, and mosquito repellant. This versatility led Malaysian islanders to name the coconut palm "tree of a thousand uses," and in parts of the Philippines it's simply "the tree of life." But, for sheer ingenuity, nothing matches the bizarre ecology of the seed itself.

When a mature coconut drops from its mother tree, it usually hits sand. Tolerance to salt, heat, and shifting soil helps wild coconut palms thrive on the upper fringe of tropical beaches, from where high tides and storms regularly carry their seeds out to sea. Once afloat, a coconut can remain viable for at least three months, riding winds and currents for journeys of hundreds or perhaps thousands of miles. In that time, the endosperm continues to solidify, but enough coconut water remains to help the seed germinate when it finally washes up on some dry, sandy backshore. With its liquid endosperm keeping things moist inside, and the rich, oily copra providing energy, a young coconut can grow for weeks on end without any outside inputs. It's not uncommon to see sprouted coconuts for sale as nursery stock in tropical markets, their bright young leaves already several feet tall.

The coconut palm's seafaring adaptations set it apart, but still fail to explain why its seeds need such an unusually rich, oily lunch.

After all, starches or cocoa butter would float, too, if you packed them inside that giant, fibrous husk. My investigation of the almond led quickly to the same basic question. Domesticated from a Central Asian cousin of peaches, apricots, and plums, the almond tree spread first to the Mediterranean and then around the world. People appreciated both its distinctive flavor and its nutritional value, because in addition to oil, an almond seed stores over 20 percent of its energy as pure protein. But why? What drove the evolution of such diverse seed nourishment strategies? Clearly, the answer to that question lay beyond what I could see in the remains of an Almond Joy. While I don't need anyone's help in *eating* candy bars, it was now apparent that I needed help understanding their biology. I decided it was time to contact someone whose name had cropped up again and again in my research, and whom more than one expert had described as a "god" in the world of seeds.

"That question?" he said, laughing. "I always ask our doctoral students that question in their qualifying exams. So far no one has come up with the answer!"

As a professor of botany at the University of Calgary and then the University of Guelph in Canada, Derek Bewley has been stumping students with seed questions for more than forty years. Luckily for everyone, his own research has provided many of the solutions. From development to dormancy to germination, the Bewley lab has explored all aspects of seed biology. But in spite of all these scholarly accomplishments, he told me his career had come as something of a surprise.

"Green was not a color where we lived," Bewley explained, recalling his childhood in the "smoky, dirty old town" of Preston, Lancashire. "We lived in what you would call a row home. There was no yard in front, and all we had in the back was a bit of concrete before the alley started." Life might have turned out quite differently if Bewley's grandfather hadn't retired to the country, where he raised tomatoes and bred award-winning chrysanthemums and dahlias. Visiting granddad and watering those greenhouses became one of Bewley's "great joys as a child." It sparked a passion for the green things of the world, and the seeds that produce them. That passion has produced hundreds of research papers and four books, including co-editorship of the seven-pound, eight-hundred-page *Encyclopedia of Seeds*, a constant companion to me in my own research. I knew I'd called the right person, but within a few minutes I also realized I wasn't going to get a simple answer.

"The evolution of this doesn't seem to be logical," he began, and told me how starches, oils, fats, proteins, and other energy strategies seem to be scattered at random across the plant kingdom. No one technique stands out as more advanced than another, since many re-cently evolved species store energy in the same basic ways as ancient ones. To make matters worse, seeds usually contain several different kinds of energy, and a mother plant might change the proportions based on variations in rainfall, soil fertility, or other growing con-ditions. Nor do plants in similar environments or with similar life histories necessarily rely on the same strategy. Grass seeds are noto-riously starchy, but one of the most common weeds in a grain field is the annual mustard called rape, whose tiny seeds produce copious quantities of canola oil. (Like "coconut water," the name "canola" is a savvy branding invention. No one, presumably, felt very optimis-tic about marketing a product called "rape oil.")

"There is one general rule," he finally admitted. "Oil and fat-storing seeds have the most energy per weight. You get more punch from lipids than from a big pile of starch." He also told me that seeds don't usually access that energy until *after* germination. Most species keep enough sugars on hand to spark the embryo to life, and then start the more complex process of accessing their stored reserves. Starches convert to sugars relatively easily, but it takes a whole se-ries of events to change protein, fat, or oil into a form useful for cell activity. Our own bodies work the same way, which is why you see competitors in Ironman triathlons downing bananas, cereal bars, or even jam sandwiches rather than slabs of bacon or cups of olive oil. In terms of seed evolution, this puts the emphasis on the newly sprouted plant and the resources its growing conditions will demand. But while that may explain why forest seeds like cacao and almond use fats and oils to fuel slow, steady growth in the shade, it does nothing to explain why mustard seeds in wide-open fields use the very same things to grow quickly. "There are exceptions," Bewley said. We were talking on the phone, but I could almost see him shaking his head. "There are always exceptions."

The British physicist William Lawrence Bragg once said that science is less about obtaining new facts than "discovering new ways of thinking about them." Talking to Derek Bewley didn't settle my questions about seed energetics with new information. Instead, it did so by reminding me of an important and fundamental truth about evolution itself. Charles Darwin once wrote, "Man may be excused for feeling some pride at having risen . . . to the summit of the organic scale." This statement was fitting to its time, an era when any respectable Victorian gentleman naturally placed respectable Victorian gentlemen on the top rung of the evolutionary ladder. The trouble lies in the whole notion of evolutionary ladders and summits, the idea of a directional process climbing toward some notion of perfection. Of course, Darwin had a much more nuanced understanding of evolution, but this concept took root in our collective intellect and was perpetuated in cartoons, popular accounts, and even serious works of scholarship. The mind returns to it unconsciously, despite being surrounded by direct evidence to the contrary. If evolution progresses toward singularity, then how do we explain diversity—the 20,000 different grasses, the 35,000 dung beetles, the profusion of ducks, rhododendrons, hermit crabs, gnats, and warblers? Why are the most ancient life forms on the planet, bacteria and archaea, more diverse and prolific than all other species combined? Given time, evolution is much more likely to provide us with a multitude of solutions than it is to give us one ideal form.

My mistake lay in assuming that seeds had perfected the "best" methods for storing energy. I wanted to think that natural selection had eliminated the various possibilities until only one or at most

FIGURE 3.2. Darwin looks on in this cartoon parody from *Punch* magazine, December 6, 1881. Entitled "Man Is But a Worm," it shows a spiraling progression of forms, from worm to monkey to evolution's presumed pinnacle, a top-hatted Victorian gentleman. WIKIMEDIA COMMONS.

several strategies remained, each adapted to a particular environment (forest, field, desert, etc.). The reality is far more complicated and far more interesting, like evolution itself—an endless and elegant articulation of the possible. Just as seeds can pack their lunches in different places (cotyledons, endosperm, perisperm, and so on), so, too, can that energy take many forms. If they offered only starch, seeds would no doubt still be successful in nature and we would still depend on them as a staple food. But without oils, fats, waxes, proteins, and other fuels, the seed habit might have lacked the versatility to dominate so many terrestrial ecosystems. And people would not be able to rely on peas, beans, and nuts for over 45 percent of global protein consumption. Nor could we enjoy most deep-fried foods, walk on linoleum floors, paint our houses, lubricate rocket and race-car engines, or marvel at the artwork of Vermeer, Rembrandt, Renoir, van Gogh, and Monet. All of these activities rely on seed-based oils. Even the most unusual energy sources in seeds turn out to have valuable human uses. The tagua nut palms of South America pack their lunches by thickening every cell wall within the endo-sperm, sometimes to the point of squeezing out the cells' living con-tents. The resulting seeds are so hard they can be cut and polished for buttons and jewelry, carved into figurines, or used as a replacement for elephant ivory in the manufacture of chess pieces, dice, combs, letter openers, decorative handles, and fine musical instruments.

"Success is an endpoint in itself," Bewley told me. The constant iterations of evolution ensure that new seed strategies will emerge, and anything that works is likely to stick around. In an odd way, this point took me right back to Almond Joy bars, and the catchy jingle that got me hooked on them in the first place: "Sometimes you feel like a nut; sometimes you don't." The advertisements fea-tured "nutty" people eating Almond Joys while skydiving or riding horses backward, alternating with more straight-laced types eating Mounds, which is basically the same confection minus the almonds. Combined with the sort of irresistible tune that neurologist Oliver Sacks called a "brain worm," these ads propelled both Almond Joy and Mounds into the top tier of American candy sales. But they also provide an important evolutionary lesson. When the goal is to sat-isfy a sweet tooth, tweaking the contents of a good recipe can pro-vide more than one successful product. Similarly, when the goal is to nourish baby plants, many solutions are possible, and, like an inven-tive chef at a chocolate factory, evolution will eventually find them.

Before setting my Almond Joy experiment aside, I scanned the minor ingredients and noticed two more seed products worthy of mention: lecithin, from soybeans, and PGPR (polyglycerol polyrici-noleate), from castor beans. In seeds, they're both derivatives of storage fats, and lecithin plays an important role in mobilizing energy reserves. In chocolate bars, they're added for smoothness and act as emulsifiers, helping to keep particles of sugar suspended in the cocoa butter. Soy lecithin shows up in all kinds of other products as well, from margarine and frozen pizza to asphalt, ceramics, and non-stick cooking spray. It's even taken as a supplement for cardiovascular health, touted as an all-natural way to lower cholesterol.

After the emulsifiers, the list wrapped up with various preservatives, caramel coloring, and a warning about allergens, but I saw no sign of the last seed commodity I was looking for. Finding it required me to venture beyond my candy bar to one of its spin-off products: a trademarked Almond Joy Fudge-and-Coconut-Swirl, made by the Breyer's Ice Cream Company. There, alongside the skim milk and artificial flavors, was guar gum, an extract whose strange properties affect everything from the texture of ice cream and gluten-free bread to the price of a motorcycle in northern India. Perhaps no single example better illustrates the wonderful variety of energies stored in seeds, and the unexpected ways in which they touch our lives.

Guar gum comes from a scruffy-looking cluster bean grown primarily on farms in Rajasthan, India's, "Desert State." Botanists put it with the *endospermic legumes*, a small group whose seeds lack the hefty cotyledons we know from beans, peanuts, and other members of the pea family. Instead, guar seeds store their energy in an endosperm loaded with highly branched carbohydrates. The diagrams in a chemistry textbook make those molecules look like maps of the London Underground, but to a baby guar plant in the Rajasthan Desert, they are a simple and essential adaptation.

"These tissues play a dual role," Derek Bewley told me. "First, they can be broken down into food, the glucose that fuels plant growth. But they also provide a protective, moist layer surrounding the embryo." He explained how the branched molecules inside a guar seed have an incredible ability to grab water and hold on tight. For a desert plant like guar, this trick transforms every rare cloudburst into a vital germination opportunity. It's a habit that has evolved several times—locust beans can do it, and so can fenugreek—but always in places where the climate is dry.

Rajasthani farmers have grown guar for thousands of years, using it as a fodder for livestock and occasionally cooking up the green pods as a vegetable. But their fortunes began to change when people realized that guar-seed gum makes a palatable thickener eight times as effective as starch. Extracted and purified, guar gum soon found its way into everything from my Almond Joy ice cream to ketchup, yogurt, and instant oatmeal. By the year 2000, India's guar exports to the food industry topped $280 million, but that was nothing compared to the boom that lay ahead.

The term *fracking* refers to an oil and natural gas extraction process known in the industry by its full name, hydraulic fracturing. It involves drilling boreholes deep into bedrock and using pressurized fluids to break apart and hold open gas-rich seams. When the fracked well is pumped out, the valuable hydrocarbons come along for the ride. Over the past decade, this once-obscure technology has grown into a multibillion-dollar global enterprise, opening up vast new deposits of shale gas and coal-bed methane. Economists expect it to effectively end North American reliance on foreign oil, fundamentally altering the world energy market. Drillers in the United States alone now frack an estimated 35,000 wells annually. And into each one of those wells they pump several million gallons of fracking fluid, a goopy combination of water, sand, acid, and chemicals all held together by one thing: guar gum.

In Rajasthan, the wholesale price for guar has risen by more than 1,500 percent in just a few years, sometimes doubling on a weekly basis. Subsistence farmers who once fed the stuff to their cows suddenly found they could sell it for enough to buy a television, then a motorcycle. Now, many are building new houses or taking family vacations abroad. Shortages of beans in 2011 and 2012 caused several drilling operations in North America to shut down, and the stock price of oil giant Halliburton Corporation fell by nearly 10 percent the week it warned shareholders that guar prices now accounted for nearly a third of its fracking costs and would "impact the company's second quarter margins more than anticipated." Tight supplies and the soaring price tag have forced many in the food industry to look elsewhere for thickeners. Not surprisingly, they're finding alternatives in the seeds of other dry-country "endospermic" beans, including carob (from a Mediterranean locust tree), tara (from a coastal Peruvian shrub), and cassia (from the Chinese sicklepod). The fortunes of all three species—and their growers—are expected to surge on the coattails of the guar boom.

It's doubtful that any oracle could have foreseen the great fortunes waiting to be made from grinding up guar seeds and pumping them underground. Indian crop reports as late as 2007 do not even list hydraulic fracturing as a potential market. The guar story shows how innovations in the evolution of seeds can drive innovations in their use. From a guar bean's ability to retain water, we derive an industrial thickener, and suddenly seed energy is being used to extract fossil energy. For the oil industry, it represents something of a homecoming, since one of the most productive fracking sites in the world lies in the state of Pennsylvania, where the first commercially successful oil well was drilled in 1859. For seeds, drilling beneath Pennsylvania's hilly countryside marks a far more ancient return.

If the goal of hydraulic fracturing were fossils instead of hydrocarbons, the wells tapping Pennsylvania's Marcellus Shale would spout geysers of tiny snails and clamshells. They would not, however, produce a single seed. Because not only did those rocks form in a seabed devoid of plant life, they come from a time millions of years before seeds even evolved. Like any other new adaptation, seeds began as an oddity, bit players in a much larger drama. They appeared in the first years of the Carboniferous Period (360–286 million years ago), a time when most plants reproduced by spores. We know those spore plants best for what they left behind: vast swamp forests that fossilized as a shiny, black rock called coal. In Pennsylvania, coal deposits lie in the younger rocks directly on top of the shale, forming a layer so thick that it helped to fuel America's Industrial Revolution and inspired geologists to name an entire period "the Pennsylvanian" in its honor. To glimpse the evolution of seeds, a fracker would simply need to drill shallower wells and start poking through the tailings.

Miners have always known they lived in a world of fossils, but scientists are starting to catch on, too. Recently, teams of paleobotanists— experts in fossil plants—have begun exploring and mapping old mine shafts, redefining our understanding of how and where seeds evolved. They've realized that the best way to understand a Carboniferous ecosystem is to walk through one, and the only place to do that is in a coal mine.

Seeds Unite

*Scientific principles and laws do not lie
on the surface of nature. They are hidden,
and must be wrested from nature by an
active and elaborate technique of inquiry.*

—John Dewey,
Reconstruction in Philosophy (1920)

CHAPTER FOUR

What the
Spike Moss Knows

*The enormous quantity of vegetable debris necessary for the
formation of even a single coal bed has led to the belief that
the vegetation of Carboniferous times was ranker and more
luxuriant than at any other time in the earth's history, and
that it grew in enormous swamps under torrid cloudy climate
conditions.*

—Edward Wilbur Berry, *Paleobotany* (1920)

"It's going to be pretty much impossible to get you into a coal
mine," said Bill DiMichele, telling me precisely what I did not
want to hear. "Coal companies have been stigmatized by the double whammy of safety regulations and getting the blame for global
warming," he explained. They didn't welcome new faces on his
team, particularly not nosey, book-writing biologists.

This dashed my hopes of strolling through a Carboniferous forest,
but I couldn't exactly second-guess Bill's judgment. As the curator of
fossil plants for the Smithsonian, he'd been leading coal-mine expe-
ditions for years. Together with colleagues from various universities
and government agencies, Bill had discovered an ancient river valley

in Illinois, 100 miles long, where every detail of the forest was beautifully preserved in the rocky ceiling of the mine. "We simply look up and map the plants," he told me. "See what was growing where." He made it sound easy, but the forest emerging from those maps was anything but simple. In fact, it was redefining the whole context of seed evolution. The good news for me, he went on, was that there were plenty of places to see some of the same fossils on the surface. "Tell me what you have in mind," he said, "and I'll ask around."

Six months later, I stood next to Bill at the bottom of a desert canyon, watching dozens of paleontologists from around the world scramble up the slope toward a dark seam in the rock. "This is only a coal bed to someone from New Mexico," he said with a smile. But while it couldn't match his Illinois mine in scale, the thin vein exposed on the wall above was otherwise remarkably similar: the carbonized remains of an ancient swamp forest, with beautiful examples of its plant life preserved in the surrounding rocks.

Soon the canyon echoed with hammers ringing on stone as people reached the coal and dug in. It was the first day of a conference devoted to what paleontologists call the Carboniferous/Permian Transition, a critical time in earth's history when the climate abruptly changed from hot and humid to dry and variable. Traditionally, experts considered this a moment of triumph for seeds. The giant horsetails and other spore plants that dominated Carboniferous swamps relied on a warm, wet environment. They couldn't adapt to the changing climate of the Permian, giving seed plants the opportunity to proliferate, overcome the spore plants, and dominate the global flora. It's a nice story, but to Bill and a growing number of other specialists, there's just one problem: it's dead wrong. No one denies that spore plants declined in the Permian, but the real triumph of seeds probably came much, much earlier.

"I used to go to the field expecting certain things," he told me, explaining how textbook knowledge can burden the mind with preconceived notions. "Now I go to the field looking. I've found it's more productive to just dig a hole and see what I find." In thirty years as a Smithsonian paleontologist, Bill DiMichele has dug a lot of holes. Compact and fit in his khaki vest and baseball cap, he moved around the dig site in New Mexico with the efficiency of experience, rarely swinging a hammer, but always there to comment on a new find. "You guys are in it, man," I heard him shout at one point. "You're in it!" Bill maintains the enthusiasm of a much younger scientist, but after a few hours of conversation I understood what really lay behind his long career: insatiable curiosity. For every question I asked, he seemed to have dozens of his own. They came out in a torrent, full of fresh ideas designed to wash away layers of old thinking. Just like a paleontologist in the field, he makes his intellectual discoveries by moving a lot of rock.

This approach opened Bill's eyes to the glimmer of something new in that Illinois coal mine. Most of it looked like a typical Carboniferous forest, dominated by tree-sized spore plants related to modern horsetails and club mosses. But whenever the ancient terrain climbed upward, even a little bit, he and his colleagues saw more fossil seed plants. And when they encountered a side channel filled with debris from further up the slope, it was a jumble of conifers. No one doubts the dominance of spore plants in coal forests, but only a minor part of the Carboniferous landscape was swampy. What was growing in the uplands, on the hillsides, in the mountains?

"Hey Bill!" someone called, and motioned us over to a slab of rock at the base of the slope. There, etched in stone, was a good summary of the story I'd come to New Mexico to see. "Nice one, Scott," Bill said, as he leaned in for a closer look. (Though people had come to the conference from as far afield as China, Russia, Brazil, Uruguay, and the Czech Republic, the world of Carboniferous/Permian specialists is a small one, and they all seemed to be on a first-name basis.) The rock had split neatly down the center, revealing mirror images of two plant stems lying side by side—a giant horsetail in the genus *Calamites*, and an early seed-bearer called a *pteridosperm*, or "seed fern." The calamites stood out in sharp relief, its dark ridges

FIGURE 4.1. These fossils from a New Mexico coal bed sum up the ancient struggle between spores and seeds. They show the stem of a giant horsetail named *Calamites* right alongside that of an early seed fern. The plants grew side by side in the great wetland forests of the Carboniferous. PHOTO © 2013 BY THOR HANSON.

and grooves like a scaled-up stalk of any modern horsetail. The seed fern's trunk looked like lizard skin, scaly black and orange against the tan surface of the rock. Both species were long extinct, but for me, seeing them together embodied that ancient struggle between spores and seeds.

I snapped a picture and then scrabbled up the hillside to join in the hunt. The rock face above the coal broke apart easily, and soon I was finding my own fossils—a few ferns and horsetails, but mostly an unrecognizable jumble of leaves, stems, and spiky twigs. Around me, the paleontologists worked and talked excitedly. Where my eyes saw only dust and confusion, I knew theirs were reconstructing an ancient world. I tried to picture the calamites and seed ferns as living plants and my mind immediately turned to textbook images of the

Carboniferous: a steamy swamp festooned with huge, mossy-topped trees like something from a Dr. Seuss book, and populated by newt-like amphibians as large as horses. It was an era long before dinosaurs, let alone more familiar creatures like mammals and birds. There would have been dragonflies and a few spiders, but no ants, beetles, bumblebees, or flies. While a swamp without mosquitoes sounds appealing, the forest would have seemed strange for all that it lacked. Then I reminded myself that, if Bill was right, the landscapes of the Carboniferous might actually have looked a lot more like home.

"It should be called the Coniferous!" he burst out during one of our conversations. "The evidence now really suggests that coal was a minor element." Once Bill's team began questioning conventional wisdom, they started seeing compelling signs of a hidden flora, a community of conifers and other seed plants that lived uphill from

FIGURE 4.2. This classic view of a Carboniferous coal forest shows a swampy world dominated by ferns, horsetails, and other spore plants. Evidence now suggests that only a small part of the world was wet and hot, and that conifers and other seed plants dominated large swathes of upland habitats. Anonymous (*Our Native Ferns and Their Allies*, 1894). ILLUSTRATION BY ALICE PRICKETT, TECHNICAL ADVISER TOM PHILLIPS, UNIVERSITY OF ILLINOIS, URBANA-CHAMPAIGN.

the swamps. Though it probably covered all but the wettest places, this forest left almost no trace of itself behind—just the occasional leaves and branches that washed down from above. "There's a problem with terrestrial plants," he explained. "They don't preserve well in place." Making good fossils requires fine-grained sediments and water, common ingredients in the swamps where spore plants ruled, but rare everywhere else. So while giant horsetails and club mosses may dominate the Carboniferous fossil record, that doesn't mean they dominated the Carboniferous.

New climate research makes the case even stronger, refuting the stereotypical image of the Carboniferous as a monotonous era of hot, humid weather. Instead, it repeatedly swung from sultry periods to ice ages and back again. Coal accumulated only at the wettest times, and the wet times were interrupted by long dry spells when seed plants would have covered even more of the landscape. In this view, spore plants fall from a position of prominence to that of a relative anomaly—minor players in both geography and duration. But because they grew in swamps, they left behind an overwhelming, disproportionate, and ultimately misleading number of fossils— what paleontologists call a *preservation bias*.

"Where's Thor?" I heard someone call out. "The Czechs found some seeds!" I'd only been with the group for half a day, but everyone already seemed to know what I was working on, and just showing up had earned me a place in the first-name club. The trip leader walked over and handed me a small block of stone speckled with black marks. Through my hand lens they looked like watermelon pips ringed by thin membranes. I asked Bill what they were, but he only shrugged: "You're best off just calling them winged seeds." Fossil seeds rarely had names, he explained, because they were almost never found with the plants that produced them. Later that day I saw what he meant as I pored over trays of fossils at the New Mexico Museum of Natural History and Science in Albuquerque. There were scores of seeds, collected over decades, with labels like: "Seed?" "Ovule?" "Partial Cone?" or "Unknown Fruiting Body." In one famous case, the "seeds" of a well-known ancient plant turned out to be fossilized pieces of a millipede.

"Man, I wish someone would work on paleo seeds," a curator told me later at the conference social hour (wine, beer, and heavy hors d'oeuvres served up in a warehouse full of fossils). "We've got one that looks like a mango pit, but with a big keel like a sailboat, and it's covered with hairs. What kind of plant made that?!?"

I agreed wholeheartedly. Studying ancient seeds would open a window on Bill's hidden plant communities. After all, for every unknown seed lying in a museum somewhere, there must have been an unknown seed plant uphill from a swamp, raining its progeny down into the muck below. What's more, those seeds dated to a time when all their critical traits were just evolving—nourishment, dispersal, dormancy, defense. For seed biologists, the most exciting aspect of Bill's theory is what it does for the story of seed evolution.

The traditional view put seeds on the map at the dawn of the Carboniferous, or perhaps a bit earlier. Then, for more than 75 million years, nothing much happened. One had to accept that seed plants, with all their advantages, could only eke out a small living in the coal swamps until the climate changed in the Permian. This version of events left two glaring questions unresolved. First of all, if seeds represented such a substantial and successful evolutionary change, then why did they remain insignificant for so long? Second, if seed traits like nourishment, protection, and dormancy were so well suited to dry and seasonal climates, then how did they evolve in a swamp? Relocating seed evolution to the uplands makes these problems disappear. Suddenly, the seed strategy becomes a logical adaptation that allowed the early innovators to colonize huge swathes of unoccupied habitat. Bill and a growing number of his colleagues now think that seed plants dominated the Carboniferous, spreading and multiplying into diverse forms that the fossil record only hints at. The "rapid" rise of seed plants in the Permian finally makes sense. When the climate dried out for good, seed plants took over quickly for a very good reason: they were already there.

"I really put the pieces of this together over a long career," Bill told me, making a point of crediting his many collaborators. But overturning long-held beliefs in science never comes without controversy. "There are colleagues of mine who violently disagree with me," he admitted. "But I just try to be kind and keep smiling, keep saying it. My thesis adviser always told me, 'Don't argue; just keep working.'" Bill seems to have taken that advice to heart. After the field trip, the conference moved indoors, where people gave presentations on their research. Heated debates often erupted, but Bill always stayed out of them (and often did have a smile on his face). Later, however, I heard him restate his philosophy with a slightly different twist: "Never argue with a fool—an onlooker can't tell the difference."

If any of Bill's colleagues really do "violently disagree" with him, I didn't meet them in Albuquerque. Everyone I spoke with at the conference endorsed the notion of a dynamic Carboniferous climate, where coal forests were an interesting but by no means dominant part of the landscape. An affable Brit named Howard Falcon-Lang proposed moving back the origin of conifers by tens of millions of years, strengthening the notion of rapid seed plant evolution in the uplands. There was a Canadian graduate student who said his adviser had instructed him to "get close to Bill, and learn everything I can." But it was Stanislav Opluštil from Prague who put it best. He had once believed strongly in the traditional view, he told me, but now considered the matter settled. "Bill changed my mind."

I left New Mexico with a completely revised mental image of the Carboniferous. The big newts and dragonflies remained, but now I pictured them against a backdrop that looked a lot more like home: a forest of conifers. Bill DiMichele's work brings the story of seed evolution out of the swamp, putting it in a dry, upland context where a seed's many adaptations for aridity make sense. But it's still a long journey from spore to seed. To truly understand that leap, we must ask some indelicate questions about the private lives of plants.

When spore plants have sex, they usually do it in dark, wet places, and quite often with themselves. A fern, for example, casts off thousands or even millions of spores every year, microscopic blips that float like earthy smoke from the edges and undersides of its leaves. Each spore consists of a single thick-walled cell with no additional protection or stored energy. It will only sprout if it lands on just the right patch of damp soil, and even then it does not grow into another fern as we know it. Instead, fern spores produce an entirely separate and unrecognizable plant, a tiny, heart-shaped nub of green smaller than a fingernail. It is this plant, the *gametophyte*, that has the equipment necessary for fern sex.

When gametophytes make eggs, they also send forth swimming sperm that can paddle their way through muddy water in the soil for an inch or two (two to five centimeters). Only if that journey unites the sperm with a nearby egg will fertilization take place and a new, familiar-looking fern sprout up. The details of this system vary, but all spore plants relegate sex to a separate generation, and all of them require water for their sperm to find an egg. Those traits work fine in wet weather, but they were problematic whenever the great swamps of the Carboniferous started to dry up. Reproduction became a challenge, and having two stages to their life cycle made it doubly hard to adjust to the changing climate.

"If spore plants wanted to make major adaptations," Bill explained, "then both phases of their life cycle had to adapt to it. And that's very difficult." In other words, the tiny gametophyte not only looks different, it might also have very different requirements for soil, moisture, light, or other conditions. "I used to tell my students, 'Imagine that your sperm or eggs grew up into little one-third versions of yourself, and then those little yous had to have sex to produce another you. And what if they looked different? What if they were totally independent and had no knowledge of your existence? What if you decided you wanted to live somewhere different? If they wouldn't or couldn't go there, then neither could you!'"

FIGURE 4.3. Wallace's spike moss (*Selaginella wallacei*). Like the common ancestor of all seed plants, this spike moss has taken the evolutionary leap of separating male and female spores. The males, precursors to pollen, are pictured on the upper right, emerging from their pouch like a smear of dust. The much larger female spores appear directly below. ILLUSTRATION © 2014 BY SUZANNE OLIVE.

In some ways, it's as if seeds evolved in response to the limitations of spores. Instead of banishing sex to the soil, they united parental genes on the mother plant, equipped that progeny with food, and dispersed it in a durable, protective case that could withstand the elements and sprout when conditions were right. Eventually, they even replaced the swimming sperm with pollen, eliminating the need for water. With so few ancient seed fossils to look at, experts still argue about the details of this transition. But everyone agrees that it was well underway by the early Carboniferous. And while every step may not be preserved in stone, living examples survive in the modern descendants of spore plants that persist, and even thrive, all around us. I didn't need to travel to a conference to see them—they grow right in my own backyard.

Every day, my short walk to the Raccoon Shack leads me past spore plants, from moss in the lawn to a patch of bracken fern that has survived years of mowing, weed-whacking, slash fires, and the depredations of our chickens. But the particular spore plant I wanted to see grew a few miles down the road, on a rocky bluff over-looking the sea where most visitors to our island gathered to watch for orca whales. On a clear January morning, I packed a sandwich and headed there to search for a somewhat smaller, but no less re-markable, species, Wallace's spike moss.

Calm water stretched away below in glints and ripples of tide as I walked down the short path. I couldn't help stopping for an early lunch and found an open place where I could soak up as much warmth as possible from the winter sunlight, a rarity in our neck of the woods. Before I'd even unwrapped my sandwich, however, I spotted the object of my quest, peeking out from a crack in the rock beside me. To be honest, I knew I'd find it without much effort. I often led field trips to this site, and had once looked on proudly as the majority of my botany students ignored a passing group of orcas to focus on these tiny plants. (Having grown up on the island, they were familiar with whales, but this was their first spike moss!)

I knelt down for a closer look. Spike mosses trace their ances-try straight back to the giants of the coal forests. While this plant grew only a few inches tall, the leaves pressed to its little stem would have looked right at home on the fossils I'd seen in New Mexico. But what the spike moss knows sets it apart from almost every other spore plant that has ever lived. I pinched off the tip of a branch and held it up in the sunlight, squinting hard. Then I rubbed my eyes and sighed. It was time to admit that I'd reached a point in life where I could no longer enjoy the pleasures of spore viewing if I for-got to bring my reading glasses.

What I was looking for came into clear focus back at the Raccoon Shack, with the help of a dissecting microscope. There the spores practically glowed, tucked into speckled golden pouches at the base of each leaf. But it's not unusual for tiny things to look beautiful under magnification. What made these spores remarkable was their size, or rather, their sizes. Low down on the stem each spore looked bulky and smooth-edged, like a big river stone, but near the branch tips they were minuscule, spilling out of their golden sacs like smears of reddish dust. The spike mosses know something that ancestral seed plants also had to learn: how to separate the sexes. The large spores are female, the precursors to eggs, and the small ones are male, the beginnings of sperm. This system not only increases genetic mixing, it allows the plants to start "packing a lunch," investing their energy in the female spores destined to produce a new plant. While both the males and females must still travel off and grow into gametophytes, and they still require water for their swimming sperm, this clever adaptation evolved at least four times in the spore plants. And on one of those occasions, it led to seeds.

Like unearthing a perfect fossil, looking at spike mosses is a glimpse into the past. As modern ambassadors of an ancient line, their mismatched spores mirror a critical step in the evolution of seeds. With the sexes separated, it becomes much easier to imagine the rest of the story. Over time, early seed plants learned not to cast off their female spores but to cling to them, letting the eggs develop right there on the tops of their leaves. Male spores continued dispersing and with a few tweaks became windborne grains of pollen. When that pollen landed on an egg, the plant suddenly found itself in possession of all the basic elements of a seed: a fertilized baby that could be protected, provisioned, and sent off to grow directly into the next generation. This system gave seed plants immediate advantages whenever the weather turned dry. Where spores required water for their swimming sperm and moisture-loving gametophytes, seed plants could reproduce on a gust of wind. And their durable, well-stocked offspring landed in the soil prepared to wait for just the right conditions in which to germinate and grow.

The fossil record for seed evolution remains hazy, but as with spike mosses, other modern plants help fill in the gaps. Most people recognize the ginkgo tree as a popular ornamental, or the source of herbal elixirs sold to boost memory and improve blood flow. But it's also the sole survivor of an early seed-plant family whose pollen still produce swimming sperm, a holdover from the spore era. A group of palm-like trees called cycads also retain this trait, and one of them boasts sperm so huge they can be seen by the naked eye. (Festooned with thousands of waving tails, the sperm of the *chigua* from coastal Colombia exceed those of any other plant or animal.) Together with the conifers and a handful of lesser-known species, these plants make up the *gymnosperms*, or "naked seeds," so named because their seeds mature unadorned on the surface of leaves or cone scales.

Gymnosperms dominated the world's flora from the dry periods of the Carboniferous all the way through the time of the dinosaurs, and they remain extremely common today. Anyone who has enjoyed pine nuts on a plate of pesto is familiar with naked seeds. So are the billion or so people who live in or around temperate forests, where pines, firs, hemlocks, spruce, cedar, cypress, kauri, and other conifers still cover more land area than any other plants. But while they may be widespread, these venerable trees and shrubs long ago passed the crown of plant diversity down to a younger group of seed innovators.

The final major step in seed evolution occurred when a few gymnosperms learned to cover up. They did it in much the same way people do after a bath, and for similar reasons. At three years old, my son Noah still uses the blue plastic tub we bought when he was an infant. He can climb out on his own now, but when he does I wrap him up immediately in a big fluffy towel. I do this not out of some prudish aversion to nudity, but because his little naked body seems so vulnerable. For me, it triggers an instinctive parental response to protect and nurture. While plants don't run around making conscious decisions about towels, the same evolutionary drive led one line of gymnosperms to wrap their naked seeds, folding up the underlying leaf to enclose the developing egg. Botanists call this chamber the *carpel* and the plants that have one are known as *angiosperms*, Latin for "seeds in a vessel."

I didn't see any fossil angiosperms in New Mexico. "Wrong conference," one attendee told me gruffly. The rocks were wrong, too, off by several major geologic time periods. While it sounds like a simple, even obvious step to wrap a protective leaf around a seed, angiosperms didn't work it out until the early Cretaceous, after naked seeds had been commonplace for more than 160 million years. To put that into perspective, the entire diversity of placental mammals, from rodents and bats to whales, aardvarks, and monkeys, has evolved in a time period less than a third as long. Botanists still puzzle over this delay, but no one disputes that putting seeds in a vessel turned out to be a good idea. Once established, angiosperms spread so fast that Darwin considered their rise an "abominable mystery" that threatened his concept of measured, incremental change. They now make up the vast majority of all plant life, and their seeds dominate the discussions in this book.

From an evolutionary standpoint, the leap from spores to gymnosperms was the paramount step for seeds. Bill DiMichele laments our tendency to focus on angiosperms. "It misses the story," he told me. "There just happen to be a lot of them." But there's no doubt that wrapping those naked seeds refined the system and opened a range of new opportunities. After all, a towel is just the beginning. Noah's wardrobe trends toward striped pajamas, but people can cover up their nakedness with whatever they want: shorts and a Hawaiian shirt, a cocktail dress, or even a suit of armor. Seed coverings soon evolved from simple leaf tissue into the dizzying array of structures we know collectively as fruit. Like clothing, fruit can be protective but it can also attract, giving angiosperms a powerful way to hoodwink animals into dispersing their babies. (We will explore the ties that bind fruits, seeds, and animals, including people, in Chapter 12.)

Even more important than the evolution of fruit, however, was the way that covering seeds affected pollination. With the egg hidden away inside its vessel, wind became a less reliable tool for pollen delivery. Instead, angiosperms turned increasingly to animals, and particularly insects, to move pollen from flower to flower. Colorful petals, nectar, fragrance—all the allure we associate with flowers—developed in response to this need, transforming pollination from a random wind splatter into one of nature's most precise (and beautiful) methods for mixing genes. This system helped propel the rapid diversification that so mystified Darwin, and it also gave rise to another name for angiosperms: "the flowering plants."

In nature, the flowering plants put sex, seeds, and dispersal on full display, spurring not only their own evolution but also that of the animals and insects with which they became so entwined. In most cases, the diversity of dispersers, consumers, parasites—and, most especially, pollinators—rose right alongside that of the plants they depended upon. But the evolution of floral sex has also proved vital to people. Without the ability to manipulate pollination and save the result as a durable seed, it's hard to imagine our ancestors ever succeeding in agriculture. Author and food activist Michael Pollan takes the case a step further, calling the practice of plant breeding "a series of experiments in coevolution" that has changed both plants and people forever. Pollan has argued that human desires for sweetness, nourishment, beauty, or even intoxication have become encoded in the genetics of our crops. Selecting for these traits both pleases us and benefits the plants as we dutifully disperse them from their original habitats to gardens and farm fields across the globe. But our intimacy with seed plants fills more than our bellies—it also feeds the human imagination. The knowledge we've gained from this long relationship may be our deepest reservoir of insight into the workings of nature. Without it, the most famous experiment in history might never have taken place.

CHAPTER FIVE

Mendel's Spores

The various forms of Peas selected for crossing showed differences in length and color of the stem; in the size and form of the leaves; in the position, color, size of the flowers; in the length of the flower stalk; in the color, form, and size of the pods; in the form and size of the seeds...

—Gregor Mendel,
Experiments in Plant Hybridization (1866)

"**P**lant the peas on President's Day." Living with an avid gardener, I had come to know this adage as both mantra and command. For Eliza, sowing peas marked the much-anticipated start of a new season, and the soil in her garden was always fresh-turned and waiting well in advance. This year we both had plans for peas, but as I yanked another clump of grass from the Raccoon Shack's overgrown flowerbed, it was quite clear that mine would be late. Considering that I still had to bring in fresh soil and devise some kind of protection from the chickens, let alone order the seeds, I'd be lucky to get them in by Palm Sunday. Still, that schedule probably put me closer to planting day in Brünn, Moravia (now Brno, Czech Republic), where the famous garden I hoped to evoke was still locked away under snow.

Weather aside, Gregor Mendel had a lot of things going for him when he coaxed his first row of pea shoots to life in the spring of 1856. His abbot, Cyrill Napp, ran the Augustinian Monastery of St. Thomas more like a research university than a cloister, encouraging his monks in their studies of everything from botany and astronomy to folk music, linguistics, and philosophy. They enjoyed good food, an excellent library, and ample time to pursue their research. For Mendel, the abbot went so far as to build a dedicated greenhouse and turn over use of the orangery and a wide swathe of the monastery gardens. But the young monk also benefited from millions of years of evolution, because without the unique characteristics of seeds, making his famous discoveries would have been a lot more challenging, if not impossible.

Try to picture the father of modern genetics conducting his experiments on spore plants. He would have spent every day on his hands and knees in the mud, searching for tiny gametophytes and hopelessly trying to corral their sperm and eggs. How could one possibly control the breeding of plants whose sex happens out of sight in the soil, with microscopic, free-swimming sperm? Spore plants simply don't lend themselves to manipulation, which is why the handful of ferns and mosses that have ever been domesticated remain essentially unchanged from their wild ancestors. (It's worth noting another spore trait that prevents them from being particularly useful to people: because they don't "pack a lunch" for their babies, spores have no nutritional value. People might nibble on the occasional leaf of a spore plant, but—with very few exceptions—you can't make bread, porridge, or anything else from the spores themselves.)

Mendel never considered studying ferns or mosses. As a farmer's son, he knew enough about plants to realize that such a haphazard mating scheme could never teach him about heredity. But he did try his hand with mice, reportedly stopping only after the local bishop found it unseemly for a monk's quarters to be filled with cages of rapidly multiplying rodents. When he finally settled on peas, Mendel found a system ideally suited to his experiments. Hand-pollinating the flowers allowed him to play matchmaker, selecting exactly which plants he wanted to cross and then watching how their traits were passed down. Unlike spores, the seeds of his pea plants united genes from both parents into something that could be easily sorted, examined, and counted. And unlike mice, they lived outdoors, smelled sweet, and even provided a tasty surplus for the monastery kitchen.

When my seeds arrived, I opened the packages immediately and dumped a few of each kind onto the kitchen table. They were all the same species, but varied just as Mendel's had—some were green and speckled, some brownish; some were wrinkled, and some smooth. In nineteenth-century Moravia, Mendel had no trouble purchasing thirty-four pea varieties from his local seed vendors. I only had room to plant two, but with a little research I'd tracked down one type that Mendel himself might well have grown. The Württembergische Wintererbse, or "Württemberg winter pea," gets its name from a former kingdom in what is now southern Germany. Rail lines connected Württemberg to nearby Moravia, and the two regions were on friendly terms when Mendel was pea shopping. They even fought on the same side during the Austro-Prussian War in 1866, the same year he published his results. While the idea of a monk puttering around in the monastery garden seems inherently peaceful, Mendel lived during tumultuous times. Europe's aging empires strained under the weight of popular unrest and shifting political alliances, while scholars struggled with an equally pressing intellectual upheaval: the theory of evolution by natural selection.

The first printing of Charles Darwin's *Origin of Species* sold out in a single day in 1859, and a German translation appeared within a year. Mendel's copious notes in the monastery's copy show that he was well aware of its contents as he toiled away on his peas. But whether or not he fully grasped the significance of what those peas were telling him remains a matter of debate. Though in hindsight he seems a genius, Mendel never achieved fame during his lifetime, and very little is known about what he actually thought. (In what must be considered one of history's most unfortunate housekeeping incidents, a subsequent abbot at the monastery had all of Mendel's notebooks and papers burned.) We do know that he was far more than a dabbler. His meticulous methods and statistical approach to science were decades ahead of their time. And he applied them not only to peas, but also to thistles, hawkweeds, and honeybees, suggesting that he really was interested in general laws of inheritance. We also know that Mendel felt he'd discovered something important. Though his paper appeared in a relatively obscure Moravian journal, he ordered forty reprints and sent them to many of the leading scientists of the day. Several of those copies have since been rediscovered, unopened and unread.

Sunset's *Western Garden Book* recommends planting peas an inch deep and spacing them two to four inches apart. "Plant them closer," Eliza told me. "You're bound to lose some to slugs." In Moravia, gardeners face a whole range of pea pests, from slugs and snails to weevils, aphids, and the occasional marauding sparrow. Mendel no doubt overplanted his rows, too, which means he must have sown a staggering number of peas to produce the "more than 10,000 plants" he recalled examining over the course of his study. My small effort would never compete with those numbers, but I took comfort in knowing that I could have told that monk a thing or two about *picking* peas. As it happens, my one experience in commercial agriculture came at the helm of a seventeen-ton pea-harvesting combine. I drove the night shift for the entire summer after my senior year in high school, filling dump truck after dump truck with peas from six p.m. until six a.m., seven days a week. It was a slow machine, and I spent most of my time reading novels by flashlight, but if I'd had Mendel's patience and motivation I would have been up in the hopper, counting those peas and recording every subtlety of color, shape, and size.

Revisiting Mendel's work taught me a lot more than how many peas to plant in a row. His experiments show how profoundly seeds, and our intimate relationship with them, have influenced the way we understand the natural world. They sparked an insight into the

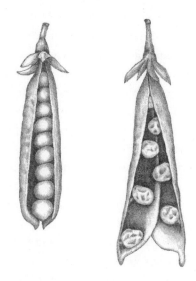

FIGURE 5.1. Common pea (*Pisum sativum*). The common pea made a perfect study species for Gregor Mendel because it shows a range of easily manipulated features, including two forms of seeds, smooth and wrinkled. ILLUSTRATION © 2014 BY SUZANNE OLIVE.

process of evolution that was quite different from the work of Charles Darwin, though just as meaningful.

It's no coincidence that both Darwin and Alfred Russel Wallace, co-discoverer of natural selection, had their epiphanies while traveling in far-flung places—Darwin on the *Beagle*'s long voyage, and Wallace in the Malay Archipelago. Grasping such a broad law of nature required a broad view of nature. The novelty of seeing exotic creatures spread across unknown landscapes helped both men discern patterns of life that can be obscured by familiarity. When you see it in the backyard, a finch is just a finch. But to understand the nuts and bolts of evolution, how individual traits are actually passed from one generation to the next, required a focus much closer to home. Mendel's revelation came by reexamining a natural system that people know better than any other. Though he never took to the farming life, he used techniques perfected by countless gardeners and farmers before him, transforming the most basic insights of agriculture into scientific laws of heredity.

When archaeologists sift through the dirt of an early settlement, they look for seeds to help pinpoint the advent of farming. If they encounter ancient grains or nuts that suddenly appear larger than the wild types, they know that someone had begun choosing plants with favorable traits. For a farmer, it's the most natural thing in the world. I once spent an afternoon with Noah, stripping dry kernels of corn from their cobs and dropping them into a metal bowl—*kuplink, kuplank, kuplunk*. We planned on grinding them all to make cornmeal mush, but if we'd been looking for seeds to save, the choice would have been obvious. Among all those tough old ears we found one with large, fat kernels that fell from the cob with ease. Big grains, easy to process—just the kind of traits to pass on.

By Mendel's time, plant breeding had progressed to a point where every region boasted dozens of local varieties of peas, not to mention beans, lettuce, strawberries, carrots, wheat, tomatoes, and scores of other crops. People may not have known about genetics, but everyone understood that plants (and animals) could be changed dramatically through selective breeding. A single species of weedy coastal mustard, for example, eventually gave rise to more than half a dozen familiar European vegetables. Farmers interested in tasty leaves turned it into cabbages, collard greens, and kale. Selecting plants with edible side buds and flower shoots produced Brussels sprouts, cauliflower, and broccoli, while nurturing a fattened stem produced kohlrabi. In some cases, improving a crop was as simple as saving the largest seeds, but other situations required real sophistication. Assyrians began meticulously hand-pollinating date palms more than 4,000 years ago, and as early as the Shang Dynasty (1766–1122 BC), Chinese winemakers had perfected a strain of millet that required protection from cross-pollination. Perhaps no culture better expresses the instinctive link between growing plants and studying them than the Mende people of Sierra Leone, whose verb for "experiment" comes from the phrase "trying out new rice."

Unlike the countless plant breeders who came before him, Mendel was not content to manipulate a system he didn't understand. His genius lay in curiosity, patience, persistence, and a considerable knack for math. Over the course of eight years, he carefully bred his peas to track the fate of specific features over many generations: stem length, pod color, flower position, and, most famously, wrinkled versus smooth seeds. By meticulously noting which parents produced what kind of offspring, he discovered that traits behaved in very predictable ways. While most of his contemporaries, including Darwin, believed that breeding led to the blending of parental types, Mendel knew that traits were passed down in discrete units. His peas taught him that every individual carries two variations of every trait, one inherited at random from each parent. In modern terms, we say that individuals carry two *alleles* for every gene. Some alleles are *dominant* and always express themselves (e.g., smooth peas), while others are *recessive* and ride along invisibly unless an individual has two of them, a situation geneticists call *double-recessive* (e.g., wrinkled peas). Nowadays, these concepts sound at least vaguely familiar to anyone who has faced the Punnett Square exercise in a basic biology course. In fact, most textbooks use Mendel's peas as an example: crossing pure strains of wrinkled and smooth produces smooth offspring, but the generation after that will include both smooth and wrinkled peas in a ratio of three to one. It's a classroom exercise now, but in 1865 Gregor Mendel was the only person on the planet who understood it. After the last pea was plucked from its pod, he summarized his revolutionary findings in what must be the most famous and influential paper that, to this day, practically no one has ever read.

Within a few weeks, my peas beside the Raccoon Shack had sprouted nicely and grown beyond the reach of slugs. By June, they twined six feet up the makeshift trellis I braced against the porch, and I could see their first purple blossoms through the window behind my desk. Mendel called pea flowers "peculiar" because they kept their vital parts hidden between two narrow petals. But it was an ideal arrangement for controlling pollination, and I followed his careful instructions, stripping off the young stamens before dusting the stigma with pollen of my own choosing. Mendel accomplished all this before the invention of the Q-tip, and by doing it his way I soon learned that the inverted pouch from one flower delivers pollen perfectly to the stigma of another. I also came to know something of the peace he must have felt in his garden, because all that famous pollination took place just as mine did—on cool spring mornings, surrounded by birdsong and blossoms.

The final step in manipulating peas involves capping the flowers with little sacks to prevent contamination. I fashioned mine from paper instead of calico, but otherwise I thought my pea bed made a pretty good facsimile of that famed Moravian garden. The process also took me right back to my days studying *almendros* in Central America. Because while they may live in the tropics and grow 150 feet tall, *almendro* trees also belong to the pea family and come festooned with purple flowers. And even though I didn't hand-pollinate any *almendros*, my dissertation was still a direct descendant of Mendel's experiment. He opened a window into the parentage of seeds that allowed me, 150 years later, to understand entire populations from patterns in their genes—which trees were breeding, how far their pollen had traveled, and who was moving their seeds from place to place. The tools of modern genetics might be different, but I have no doubt that Mendel would have understood exactly what I was doing in the rainforest and what it all meant. Still, I wonder if he would have been as persistent in his pollen dabbing if he'd known the disappointments that lay ahead.

To say that Mendel's pea paper landed with a thud would be incorrect, because that analogy implies that it made any sound at all. From its publication in 1866 to the turn of the century, *Experiments in Plant Hybridization* received fewer than two dozen citations in the scientific literature. Darwin's work, by contrast, received thousands. When Mendel presented his findings to the Natural Science Society of Brünn, not a single question was asked. (A report of "lively" audience participation in the local newspaper is thought to have been penned by a friend, or perhaps by the monk himself.) During his lifetime, the few people who knew about Mendel's research either doubted it or didn't understand it, and he probably never enjoyed a single satisfying conversation about its implications. Making matters worse, he tried and failed to replicate his findings on hawkweeds, small wildflowers in the aster family that, unbeknownst to Mendel, rarely bother with pollination at all. Instead, they produce odd, clone-like seeds that display none of the bi-parental inheritance he so meticulously documented in his peas. It was an unfortunate choice that left him even more demoralized, doubtful, and discouraged. Biographers describe the young Mendel as a genial man who was beloved by his students and fond of practical jokes. But in his later years he reportedly withdrew more and more from society as well as from science. In 1878, a traveling seed dealer could not persuade the aging monk to even talk about heredity: "It is strange that when I asked Mendel about his work with the peas, he deliberately changed the subject."

Though it's impossible to know what Mendel was thinking, one anecdote suggests that he maintained confidence in his results and sensed the impact they would eventually have. Long after the monk's death in 1884, a colleague at the abbey recalled his fondness for a particular saying: "My time will surely come."

In my pea bed, the time had come to harvest. It was late summer, and the vines drooped in the heat, pods yellowed, peas ripe and dry within. Though Mendel often worked alone, he did have the help of various monastery novices, as well as a trained assistant. My staff consisted of one three-year-old, but Noah's enthusiasm for any seed project made him an eager accomplice. We pulled up the vines, sat down on the shady porch, and started to separate peas from pods. Quick as a flash, Noah snatched up a handful and popped them into his mouth. Years before, I'd made the mistake of taking a dog with me on a mammal-trapping project, only to watch him pounce upon and devour the start of a promising dataset. In this case I still got my data, because Noah spat the seeds right back out again. Unlike the sweet ones in Mama's garden, these peas had been grown to maturity, and they were hard and dry, like raw lentils. He didn't say anything, but gave me a disgusted look that recalled one of his earliest paragraphs, carefully phrased one morning after I served his breakfast: "Mama cook food. Papa cook poop."

As we made our way through the pile of pods, I found myself impressed again by Mendel's incredible patience, and the sheer size of his task in counting all those peas. Even though my harvest was small, reduced by an unexpected pest we'll return to in Chapter 8, things had gotten a little monotonous by the time we scooped out the final pea. But that sameness was also fascinating—in fact, it was the whole point. My experiment replicated the first generation of Mendel's work: crossing pure strains of two distinct varieties. I bred smooth, round Württemberg peas with an old American type called Bill Jump, whose seeds are distinctly wrinkled. If Mendel was right, then the dominant smooth genes should have completely swamped Bill Jump's wrinklies. And now, after months of tending, here was exactly the expected result: a small jar of smooth, round peas, as if the Bill Jump genes had simply disappeared. I picked up a handful and let them run through my fingers, sensing what Mendel must have felt: the satisfaction of understanding a system well enough to predict it.

When something is known, it's hard to imagine it as a mystery. But for decades after Mendel's discoveries, no one else glimpsed the secrets of inheritance the way that he had. He lived out his life in Brünn, dying in obscurity while scientists around the world still struggled to understand how traits were passed from parents to offspring. As one frustrated botanist put it in 1899: "we need no more *general* ideas about evolution. We need *particular* knowledge of the evolution of *particular* forms." As if in answer to that wish, three researchers published "rediscoveries" of Mendelian inheritance the following year, and the field of modern genetics was born.

Independently, they each replicated aspects of Mendel's work and came to similar conclusions. And independently, they all did so in the same way he had: by controlling pollination and examining the traits of seed plants: corn, poppies, wallflowers, evening primrose, and the smooth or wrinkled seeds of the common pea.

In nature, the consistent genetic mixing that goes on in seeds gives them great evolutionary potential. Where spore sex was haphazard and often self-induced, seeds combined the genes from two parents regularly and directly, using increasingly intricate flowering strategies. It's a habit that helped seed plants diversify and dominate in nearly every terrestrial habitat, and it sped the development of all the other seed traits we'll talk about in this book. For people, it allowed the breeding of crops as varied as string beans and starfruit, and it gave us one of our most profound insights into the process of evolution. But the genes in a seed would hardly be so useful without another characteristic that we tend to take for granted.

Breeding a generation of smooth peas brought me closer to Mendel, but I wanted to continue his elegant experiment for one more year, just to see that famous three-to-one ratio in the same way that he had. If the Punnett Square could be trusted, then breeding this year's crosses would produce a predictable number of double-recessive peas with the wrinkled appearance of a pure Bill Jump. I could only do this because I knew that a packet of dry peas would be just fine sitting on a shelf in the Raccoon Shack until President's Day rolled around again. In fact, those seeds would last for two years, three years, or even longer—slumbering away in a peculiar state of suspended animation. Gardeners rely on it, plant breeders rely on it, and so does the ecology of everything from peas to rainforest trees to wildflowers in an alpine meadow. But exactly how a seed can lie dormant for years or even centuries before germinating is a fundamental mystery that scientists are only just beginning to understand.

Seeds Endure

Can you find another market like this?
Where, with your one rose
you can buy hundreds of rose gardens:

Where,
for one seed
you get a whole wilderness?

—Rumi,
"The Seed Market" (c. 1273)

CHAPTER SIX

Methuselah

Wheat in plenty was laid up, ample for the needs of the beleaguered for a long time, and wine and oil in abundance, as well, all sorts of pulses and dates heaped up together.

—Flavius Josephus, describing the storehouses of Masada,
History of the Jewish Wars (c. AD 75)

The Roman general Flavius Silva arrived at the base of Masada Fortress in the winter of AD 72–73. History tells us he had a full legion of soldiers at his command, as well as thousands of slaves and camp followers. History does not preserve what he was thinking at that moment, but anyone who has ever seen Masada Fortress knows it must have been some version of, "Oh, shit."

Perched atop a 1,000-foot (320-meter) rock spire and surrounded by sheer cliffs, the stronghold boasted fortified casemate walls, watchtowers, and a well-stocked armory. It commanded sweeping views in every direction, and one of the only approaches was a steep, winding trail known ominously as "the snake path." What's more, the people defending Masada belonged to a particularly fierce group of Jewish rebels called the Sicarii, named for the wicked daggers they used to assassinate their enemies. General Silva must also have realized that while he and his army would be forced to camp in the harsh, rocky

FIGURE 6.1. This 1858 painting by Edward Lear, *Masada (or Sebbeh) on the Dead Sea*, shows the formidable approach to Masada Fortress. The ancient ramp erected by its Roman attackers is clearly visible as a ridge ascending from the right. WIKIMEDIA COMMONS.

desert that surrounded the fortress, the rebels had their choice of villas and palaces styled to the tastes of Masada's original builder, Herod the Great.

The Romans settled in for a long siege. Silva had orders to crush the Sicarii, the last holdouts of a widespread Jewish uprising known as the Great Revolt. Over the course of several months, his engineers erected an embankment that is still clearly visible, rising like a massive wave of earth up the western side of the mountain. When it was finished, Silva's soldiers marched to the top, breached the wall with a battering ram, and took the fortress by storm. At the time, this victory gave General Silva a major career boost. He served as governor of Judaea for eight years and later returned to Rome as consul, a position second only to the emperor. In retrospect, however, the Siege of Masada did quite a lot more for the cause of Jewish nationalism, for coin collectors, and for our understanding of dormancy in seeds.

When Silva's legionaries entered Masada, they expected to find dagger-wielding warriors, but were met instead with an eerie silence. Rather than surrender or risk capture, nearly 1,000 Sicarii men, women, and children had committed mass suicide. The story of their resistance and sacrifice has become a near-mythic symbol of endurance to the Jewish people. In the run-up to statehood, future leaders of Israel embraced Masada as an allegory for national unity and resolve. For decades young Israeli scouts and soldiers have hiked the snake path as a rite of passage, and Masada now ranks among the most popular tourist attractions in the country. If Silva returned today, he could take a cable car to the top, and he would find the phrase "Masada Shall Not Fall Again" emblazoned on everything from t-shirts to coffee mugs.

For coin collectors and seed experts, the defenders of Masada are remembered less for what they did than for what they left behind. Not wanting the Romans to recover anything of value, the last Sicarii moved their possessions and food supplies into a central warehouse and then set the building ablaze. As the wooden beams and rafters burned, the stone walls collapsed inward, forming a heap that would lie undisturbed for nearly 2,000 years. Archaeologists picking through the rubble in the 1960s unearthed a trove of ancient shekels that settled several nagging questions about Jewish numismatics. Not surprisingly, many of the coins featured the graceful curving leaves of the Judaean date palm, a tree whose fruit was both a local staple and a highly profitable export. Emperor Augustus was said to favor them, and vast date-palm orchards lined the Jordan River from the Sea of Galilee south to the shores of the Dead Sea. Digging deeper, the excavation team soon encountered provisions: salt, grain, olive oil, wine, pomegranates, and a generous supply of the dates themselves, so beautifully preserved that scraps of fruit still clung to the seeds.

While it makes perfect sense for the Sicarii to have stocked up on their country's most famous crop, finding dates at Masada was still a major event. Though mentioned in the Bible and the Koran, and praised for their sweetness by everyone from Theophrastus to

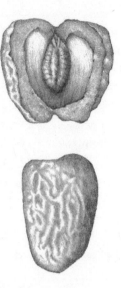

FIGURE 6.2. Date (*Phoenix dactylifera*). Cultivated since ancient times for their sweet fruits, date palms also hold the record for longevity in seeds. A date seed recovered from the ruins of Masada Fortress germinated after lying dormant for nearly 2,000 years. ILLUSTRATION © 2014 BY SUZANNE OLIVE.

Pliny the Elder, the particular date variety grown in Judaea had long since disappeared—a victim of changing climate and settlement patterns. Now, for the first time in centuries, people could see and hold the fruit that was once considered King Herod's main source of revenue. What happened next, however, was even more remarkable. Four decades after museum workers had cleaned, labeled, and cataloged the Masada dates, someone decided to plant one.

"To say I was wildly excited would be an understatement," Elaine Solowey told me, recalling the spring day in 2005 when she noticed a lone shoot poking up through the potting soil. An agricultural expert at a kibbutz in the Negev Desert, Dr. Solowey had planted "hundreds of thousands of trees" in her career before she tried the

Masada dates. "I really didn't expect anything to come up," she confessed. "I thought those seeds were as dead as doornails. Deader than doornails!" Solowey credits her collaborator, Sarah Sallon, for dreaming up the whole idea.

"It just seemed meant to be," Sallon said, when I reached her by phone. "To tell you the truth, I expected it." It was ten o'clock in Jerusalem and she'd been working late, but Sarah still launched into our conversation with enthusiasm, and somehow also managed to carry on talking with her son in the next room. She even served him a meal. Sarah's boundless energy made me wonder if the date seed hadn't sparked to life simply because she'd touched it. Trained as a pediatrician, Sallon has become a world expert on natural medicines, particularly those derived from the native plants of Israel. Her laboratory team works with Solowey's field crew to raise and test dozens of different medicinal herbs. "But I also became interested in what used to grow here," she explained, "the things that have disappeared." Ancient healers used dates from the Judaean palm to treat everything from depression and tuberculosis to common aches and pains. "Bringing it back," she mused, "might serve a greater purpose."

The sprouting palm that so surprised Elaine Solowey (but not Sarah Sallon) now stands ten feet tall and bears the name Methuselah, after the oldest character mentioned in the Hebrew Bible. But at 969 years, the biblical Methuselah had scarcely reached middle age compared to this little palm tree. Radiocarbon dating confirms that the dates from Masada had probably been stored there for decades before the fortress fell. Methuselah may look like a young tree, but its nearly 2,000-year lifespan makes it one of the oldest organisms on earth. At that age, who can begrudge it a little pampering? "We built him his own gated garden, with his own watering system, burglar alarm, and security camera," Elaine said with a laugh. "He is definitely a tree that has everything."

Elaine used the male pronoun because date palms are unisexual, and when Methuselah flowered for the first time in 2012, he turned out to have blossoms laden with pollen. To fully bring the Judaean date back from extinction, someone will have to sprout a female seed, too. When I asked Sarah if they were working on it, she seemed almost bursting with the news: "Of course we are!" she exclaimed. "But I can't tell you about it!" In science, it's never a good idea to let the cat out of the bag before all the data are analyzed, reviewed, and published. By the time this book is printed, however, Sarah and Elaine may have announced their results to the world. With any luck, those findings will tell us not only how Judaean dates live so long, but their precise flavor and sweetness, and whether or not they can cure a headache.

Methuselah's story ranks as the oldest known example of a naturally germinating seed. It's a tale of incredible endurance that provides a fitting and peaceful complement to the heroic defense of Masada, and makes it possible that Judaean dates may once again flourish in the Jordan Valley. But it's hardly the only time that an ancient seed has sprung suddenly and surprisingly to life. In 1940, the study of seed longevity received a jolt when a German bomb hit the botany department at the British Museum. After firefighters extinguished the blaze and cleared away the debris, museum workers returned to find some of their specimens sprouting. Responding to the heat and moisture, the seeds from a silk tree collected in China in 1793 had germinated and sent up perfectly normal-looking shoots. (Three of the seedlings were planted at the nearby Chelsea Physic Garden, where another bomb hit them in 1941.) Ever since then, enterprising botanists have been pushing back the record for longevity—200 years for pincushion proteas and other African exotics discovered in a cache of privateers' booty; 600 years for a canna lily seed preserved inside a Native American rattle; 1,300 years for Indian lotus seeds recovered from a dry lakebed. The most promising new developments come from the high Arctic, where a team recently transplanted live tissue from a tiny mustard frozen in a squirrel burrow for over 30,000 years. By itself the seed couldn't germinate, but the fact that any of its parts stayed viable for that long suggests that Methuselah's record is bound to fall.

"Seeds can have an almost indefinite life-span," Sarah told me, when I asked her about the limits of dormancy. Elaine's answer was more prosaic, but probably closer to the truth. All seeds die eventually, she explained, and most die within a few years or decades. But Methuselah had been found in "the perfect place"—entombed deep under a collapsed building in a bone-dry environment where it was protected from insects, rodents, moisture, and the damaging rays of the sun. During the Egyptology craze that swept through Europe and North America in the nineteenth century, people claimed that similar conditions had preserved the grains and peas buried with the pharaohs. Unscrupulous local guides made a brisk business selling "mummy wheat" to tourists, and mainstream magazines, from *Harper's* to *The Gardener's Chronicle*, touted its amazing yields and health benefits. Even today, "King Tut Peas" remain a staple offering in seed catalogs. Although there's no evidence that any of the pharaonic seed claims are true, Methuselah's story suggests that they may not be impossible.

Bringing back ancient plant varieties makes for headline-grabbing science, but it's just an extreme example of what seeds do all the time. In the broadest sense, dormancy refers to that quiet pause, however long, between when a seed matures and when it germinates. Garden seeds sold in packets are dormant, and so are the grass seeds you sprinkle in your front yard to plant a lawn: dry, hard, easy to store, and ready to sprout as soon as they hit a patch of damp soil. Without dormancy, farmers and gardeners couldn't save seed for future plantings; nor would grains or legumes or nuts last so long stored in our cupboards and pantries. We take it for granted, but if seeds couldn't lie around for months or years on end, our entire food production system would be a folly. But while the endurance of seeds is vital to people and agriculture, it's even more important to the plants themselves.

Anyone who has blown the fluff from a dandelion is familiar with the idea of seeds dispersing through space. In a very real sense, dormancy allows seeds to disperse through time. It gives plants a way to position their seeds in a particular moment, that point in the future when conditions will be just right for germination. Plants with long-lived seeds produce babies that will survive whatever winter, drought, or other barrier may stand between them and the next good growing season. They can also hedge their bets against floods, fires, or other chance events that could wipe out every seedling in a given year; dormant seeds will still be there in the soil, waiting for another opportunity. This gives seed plants an obvious evolutionary advantage anywhere that the climate is harsh, unpredictable, or strongly seasonal. It fits neatly into Bill DiMichele's theory of seeds evolving in the dry, rugged uplands of the Carboniferous, where dormancy gave seeds another clear advantage over their short-lived spore competitors. And it also helps explain why dormancy is the dominant seed strategy in nearly every environment except tropical rainforests, where favorable weather is basically constant and seeds are better off sprouting immediately to avoid greater dangers from rot, pests, and predators.

The plants that first invented dormancy probably did little more than drop their seeds early. These immature cast-offs had no special adaptations—they simply needed more time to develop before they were ready to germinate. Some species still follow a version of this approach, as any gardener who has tried to grow parsley will understand. It takes forever to sprout, because its tiny embryos must grow for days and days *inside* the seeds before they are big enough to put out a root. Over time, most plants developed the habit of holding onto their seeds longer and drying them out, reducing water content by as much as 95 percent. This remains the single most important factor in slowing down a seed's metabolism, something we'll discuss in detail in the next chapter. For now, think of desiccation as a starting point from which seed dormancy quickly evolved into an array of strategies so complex they border on the arcane. In general, and in this book, dormancy is defined broadly—any pause, for any length of time, that takes place after a seed matures and before it sprouts. But specialists like Carol Baskin make an important distinction between seeds that are simply inactive, and those that are technically dormant.

"If a seed is truly dormant and you place it on a moist substrate at favorable temperatures, it will not germinate," she told me. In other words, dormant seeds don't simply sit around waiting for rain showers and sunny days. To meet Baskin's definition of dormancy, a seed must actively resist germination, using a wide array of tricks to stave off that moment of sprouting. This sounds counterintuitive—isn't the whole point of a seed to germinate?—but it gives seeds a sophisticated way to interact with weather, daylight, soil conditions, and other factors that make up their environment. The most common tactic in temperate regions takes advantage of temperature, requiring the prolonged chilling effects of winter, followed by warming, to make the seeds ready to sprout in the spring. This strategy often works together with light requirements, which can be surprisingly specific. Some wild mustard seeds respond to changes in the angle and length of daylight through six feet of snowpack, while many forest species recognize the difference between full sunlight (a good chance to sprout), and the far-red wavelengths that filter through leaves (too shady). Whatever their needs, dormant seeds cannot and will not germinate until certain conditions are met.

"The evolution of this is driven not so much by the seed as by the seedling," Carol explained. While germination might be successful in any damp moment, the really important thing is what happens next. A mother plant's entire investment in nurturing and dispersing her seeds means nothing if they sprout in the wrong season and immediately perish from thirst, cold, heat, or shade. These high evolutionary stakes have led to the highly specific cues needed for dormant seeds to wake up. Some of the most elaborate examples come from fire-prone areas, where young plants grow best after a blaze opens up the habitat and releases a flush of ashy nutrients. Seeds adapted to this system, from particular acacias and sumac to rock rose and gorse, often remain completely waterproof and unable to imbibe until the extreme heat of open flames cracks their coats or unplugs tiny stoppers to let moisture in. Some species also require exposure to the hot gasses found in smoke, or respond to particular chemicals released from partially charred wood. Germination experts have been known to flash-heat seeds and blast them with smoke in an attempt to simulate wildfires in the lab. For desert plants, the challenge lies in differentiating occasional cloudbursts from the sustained rains that can actually nourish a thirsty seedling. Just how they do it remains controversial, but some experts believe their seed coats contain "rain gauges," chemicals that inhibit germination until they are leached away by just the right amount of water.

To Carol Baskin and her husband, no aspect of seed biology is more intriguing than how seeds sleep and what it takes to wake them up. "It just fascinates us," she told me. In all, she and Jerry have identified fifteen different classes and levels of seed dormancy, with many variations on each theme. They vary based on what causes the dormancy (e.g., an impermeable seed coat, an undeveloped embryo, a chemical or environmental constraint), and the "depth" of the dormancy (how difficult it is to overcome). From their backyard in Kentucky to the mountains of Hawaii to the cold deserts of northeastern China, they continue to turn up new wrinkles. What keeps them coming back is how little we actually understand about the process itself. Everyone agrees that desiccation is important, and scientists know many of the chemicals and genes involved, but just how a seemingly lifeless seed recognizes things as diverse as frost, smoke, heat, day length, and the ratios of wavelengths in sunlight remains mysterious. Even the basic distinction between the end of dormancy and the beginning of germination is fuzzy. In science, and life in general, it's possible to know a lot about what happens without understanding how. I know what happens when I turn on my computer, for example. I can type, search the Internet, or entertain my son's grandparents with emails and photos of his latest antics. But I don't have the foggiest notion about how a computer actually works, as my frequent calls to tech support can attest. The science of seed dormancy is more advanced than that, but there's still a great deal to learn, and that's what makes it exciting.

At the end of our conversation, I asked Carol if suspended animation was a good analogy for dormancy. (When science can't provide a complete answer, it's only natural to turn to science fiction.) "Not exactly," she responded, "because the seed is still active." This made me smile—only a seed biologist would call the hard, dry, inert little lump of a dormant seed "active." But Carol and many others believe that seeds continue metabolizing like any other living thing; they just do it very, very slowly.

When the title character opens his eyes in H. G. Wells's classic novel *The Sleeper Awakes*, he finds the world transformed. Two hundred years have passed, and everyone he has ever known is dead. On the bright side, his savings account has built up enough compound interest to make him the richest man in history. Dormancy can be a similar experience for a seed—Methuselah, after all, woke up to his own private garden. More typically, seeds sleep through a single season, or perhaps a few years or decades, but the payoff remains significant: favorable growing conditions, and with any luck, a good patch of ground. In the Wells novel, the roused Sleeper immediately meets people in strange robes who try to keep him from claiming his fortune. Seeds also awaken in a state of competition with odd bedfellows, since throughout their slumber and for many years before, other seeds of all varieties have been stacking up in the soil nearby.

Nowhere else in nature can you find a setting quite like a soil seed bank. If dormancy compares to suspended animation, then seed banks represent suspended competition: hundreds or thousands of fierce rivals, from many species and generations, all lying side by side together in wait. When the right conditions suddenly appear (particularly following a fire or other disturbance), they trigger an intense struggle to get established. This competition among so many near neighbors has been a driving force in seed evolution, influencing everything from seed size to the speed of germination to the quality and quantity of food reserves. Some experts believe that seed banks even introduce new genetic variation into plant populations, since the DNA in older seeds often begins to degrade and accumulate odd quirks. For the people who study seed banks, their surprising diversity and longevity can even inspire use of the rarest punctuation mark in science, the exclamation point. Charles Darwin once sprouted 537 seeds from three tablespoons of pond mud "all contained in a breakfast cup!"

Because they endure so long, seed banks can provide fascinating glimpses into the past. In Methuselah's case, the "bank" was an ancient storehouse, but even in natural settings the seeds preserved in soil often include species that have disappeared from the landscape. Ecologists turn to seed banks when they want clues about historical habitats—what plants used to grow where. Darwin's fascination with seeds began when he observed rose mallow, a plant of fields and gardens, sprouting where a new roadbed was being dug through a dark forest. He concluded that the seeds must have been in the soil "undisturbed for ages," remnants from a time before the trees, when the landscape was open and cultivated. The most dramatic examples of rediscovery come when soil disturbance reveals long-forgotten seed banks, sometimes in unexpected places. In the spring of 1667, Londoners stood amazed as their city burst into blossom. Fields of golden mustard and other wildflowers suddenly appeared, spreading north from the River Thames, where, six months earlier, the Great Fire had razed thousands of homes and buildings, exposing bare ground and a seed bank buried for generations.

Seed banks may give us a glimpse backward, but for plants, the whole notion of dormancy remains focused on the future—dispersing their progeny forward in time. Perhaps no one is more conscious of this than gardeners and farmers, who find themselves pulling weed sprouts from the same patches of ground year after year after year. In fact, it was a group of frustrated farmers that inspired Professor William James Beal to embark on what has become one of the longest-running scientific experiments in history. A botanist at the Michigan Agricultural College (now Michigan State University), Beal started his project in the fall of 1879 in response to an appeal from local farmers. They wanted to know how many years of pulling and cultivating would exhaust the weed seed in their fields. To find an answer, Beal carefully buried twenty glass bottles on a hill near his office. Each bottle contained fifty seeds from twenty-three local species, chosen "with the view of testing them at different times in the future." Over the next three decades, Beal dug up one bottle every five years, planted the seeds, and kept track of how many still germinated. When he retired, he handed on the experiment to a younger colleague (complete with a "treasure map" to the secret location of the buried seeds). Later caretakers extended the time frame, so that the last "Beal bottle" won't be exhumed until 2100. It's unknown whether descendants of those original farmers are still working the same fields, but if so, we know they would still be yanking up weeds like moth mullein and dwarf mallow. Seeds of both species germinated readily from a bottle dug up in 2000, after 120 years beneath the soil.

Many people now regard Beal's experiment as a novelty, a charming holdover from the era of the great nineteenth-century naturalists. His simple idea continues to remind us, every few years, that seeds can live a long, long time. But while modern research methods have grown more complex, Beal's work foreshadowed major developments in the study of seeds. Never before have scientists tucked away so many seeds for the future—billions of them, from thousands of species. But instead of glass bottles, we now store seeds in high-security vaults and frigid arctic caves. Like Beal, modern seed savers take out their specimens every so often and germinate them; but unlike the good professor, they are not trying to understand old seed banks. Instead, they are creating new ones.

Take It to the Bank

*The work is in the hands of Prof. Vavilov.... In his travels
through Turkestan, Afghanistan and neighbouring countries
and by a vast correspondence, collections of seeds of wheat,
barley, rye, millet, flax, etc., have been brought together on
a great scale. The central office is in Leningrad and occupies
a very large building, in great measure a living museum of
economic plants as represented by their seeds.*

—William Bateson, *Science in Russia* (1925)

Horsetooth Reservoir fills a six-and-a-half-mile canyon due west
of Fort Collins, Colorado. Four dams hold back the water, their
high earthen walls clearly visible from various parts of town. Should
one or more of them fail, floodwaters would reach the city center in
less than thirty minutes, too soon for any organized evacuation. A
government study concluded that all or parts of the city, as well as
several other communities downstream, "would be severely damaged
or destroyed." Reconstruction and recovery estimates top $6 billion.

There is one building, however, that is expected to do just
fine. It lies on the edge of the Colorado State University campus,
wedged between the ROTC center and a track-and-field facility.
The name on the door reads National Center for Genetic Resources

Preservation, but most people still know it by its former name: the National Seed Bank. A casual observer would never guess that its nondescript cinderblock walls house laboratories and cryogenic vaults built to withstand earthquakes, blizzards, long-term power outages, and catastrophic fires. And on the off chance that the Horsetooth dams should burst, the building is designed to float.

"There's a double foundation," Christina Walters explained as we passed through a wide interior door. "It's like a building within a building." The seed collection lies inside that central core, safe from as much as ten feet of floodwaters. "They were thinking about tornadoes, too," she added. "The walls are reinforced concrete. You couldn't hurt this place with a Cadillac going 75 miles per hour."

It's not clear why anyone would assault the National Seed Bank with a Cadillac, but I laughed at the image. I did a lot of laughing with Chris Walters. An energetic woman of middle years, she talked about seeds with a charming mix of intensity and humor, and after every joke her eyes kept smiling long after the conversation had moved on. "Let's go in," she said, and another door whooshed open in front of us. Inside, lights brightened automatically as we walked by rack after rack of long, movable shelves, the type that libraries use to save space. And with more than 2 billion specimens in the collection, space is at a premium at the National Seed Bank.

"We're part of the Department of Agriculture, so crops are definitely a focus," Chris explained. The collection includes varieties of every imaginable food plant as well as samples of their closest relatives from the wild. The idea isn't just to stockpile popular crops, but to save the range of genes that make them useful—from subtleties of flavor and nutrition to drought tolerance or resistance to disease. Seed banks store their thousands of varieties with a larger goal in mind: preserving, and better understanding, diversity itself. "What's this?" Chris asked, snatching a silver foil bag from the nearest shelf. "Ah, sorghum," she said. "I love sorghum."

It's safe to say that Chris Walters loves more about her job than the sorghum. She started at the seed bank as a postdoctoral fellow in

1986 and worked her way up to supervising the entire research program, from germination to genetics. Like Derek Bewley, she credits her passion for plants to a grandfather who had a farm. Her own family moved a lot and never even planted a garden, but she remembers begging her mother to buy her the little ornamentals sold at grocery stores. "They were just coleus plants," she said, laughing. "You know, the ones with the purple leaves!" In college, her botanical interests began to focus on seeds, but it wasn't always smooth sailing. One professor suggested she'd be better off studying "real plants." But Chris persevered, specializing in desiccation, longevity, and physiology. Thirty years later, there are few people in the world with a better understanding of just what goes on (and what doesn't) inside a dormant seed.

"I've been in here long enough," she said suddenly, putting the sorghum back and heading for the door. I was happy to follow. Seeds last a lot longer if they're cold, and massive refrigeration units keep the collection room chilled to a constant 0°F (–18°C). We exited shivering, with clouds of vapor swirling around our feet, and I now understood why the coat rack outside was draped with parkas and winter jackets. The tour continued to another vault below, where seeds were kept even colder in steel vats of liquid nitrogen. "Seeds have different personalities," Chris told me, and explained how manipulating two critical storage factors, temperature and humidity, helped them find the best fit. When they got it right, the results could be dramatic. A grain of rice might stay viable for three to five years in nature, but could live for two hundred years at the seed bank. Their wheat specimens did even better, on track to last twice that long. "There's no such thing as immortality," she qualified. "Nothing lasts forever." But seeds in a facility like the National Seed Bank come pretty darn close.

When we reached her office, I asked Chris to explain how seeds do it—how a seemingly inert object could survive for so long. Like every other expert I talked with, she immediately pointed out how little we really understand about seeds. But then she homed in on the things that scientists do know. "When a seed dries out, the enzymes slow down and the molecules stop moving," she explained, shifting piles of books and papers from two chairs so we could find a place to sit down. "Metabolic activity basically grinds to a halt." Then she produced illustrations, diagrams, and even an electron micrograph of desiccated seed cells. With the water gone, they looked like crumpled plastic sacks clumsily stuffed with lumps. If you've ever let your three-year-old bag the groceries, you've seen something similar. "It's a mess in there," Chris said, "and very hard to study because you can't see anything." But Chris's work does show that the reactions necessary for a plant cell to function, the very basics of metabolism, rely on water. Take out the water and everything stops. Put it back in, and the seed comes alive.

I asked her if a packet of dry soup mix might be a good analogy— it's just a jumble of stuff, but when you add water you end up with a tasty meal. "Yes, to a point," she said, and then frowned. "The difference is what happens when you put the water back in. Soup mix gives you soup, a bunch of ingredients floating around at random. In a seed you get organized, functioning cells. Somehow, desiccated seed cells have the ability to remember and regain their structure. That's unusual. Most cells can't do it." Then she looked across at me and the laughter was back in her eyes. "If we dried your cells out and then added water, we'd get soup."

Luckily for me, and for most members of the animal kingdom, life and reproduction don't require surviving desiccation. But there are a few creatures who have learned this trick: certain nematodes, rotifers, tardigrades, and a group of tiny crustaceans familiar to generations of comic-book readers. Though they don't actually wear crowns or lipstick like the pictures in those famous back-page ads, the brine shrimp sold as Sea-Monkeys are no less remarkable. Like seeds, their dried eggs can survive for years—in the wild or in mail-order packets—and their cells remember exactly how to reassemble themselves as soon as they land in a fishbowl. Experts now think that desiccated seeds and Sea-Monkeys have a lot in

FIGURE 7.1. A brine shrimp (*Artemia salina*), one of the few animals whose life cycle includes desiccation and dormancy similar to that found in seeds. Photo © by Hans Hillewaert/CC-BY-SA-3.0. WIKIMEDIA COMMONS.

common, preserving vital functions in a glass-like state within their cells. Medical researchers recently mimicked this system to create the first stable dry vaccines for use in places that lack refrigeration. "Desiccation was definitely the inspiration," one measles expert told me. They started with brine shrimp, he explained, but had their best results when they suspended live vaccine in *myo-insitol*, a sugar extracted from rice and nuts.

The biology of dormancy has implications for everything from pharmaceuticals to space exploration. NASA scientists study seeds to develop new storage and survival strategies for long missions. When astronauts bolted a case of basil seed to the outside of the International Space Station, the dormant little pips did just fine, germinating normally after more than a year of exposure. At the seed bank, however, most research has a more earthbound goal: keeping people fed in a rapidly changing world. Seed banks act as giant libraries of variation that farmers and plant breeders can turn to when certain crop traits are needed. After the 2004 tsunami flooded coastal rice paddies from Indonesia to Sri Lanka, seed banks quickly provided salt-tolerant varieties to replant the fields. And when the Russian wheat aphid threatened America's grain crops in the 1980s, researchers screened more than 30,000 seed-bank varieties to find the strains with natural resistance. With commercial agriculture increasingly focused on a few, mass-produced crops, seed banks provide an important hedge against disease outbreaks, natural disasters, and the steady loss of food-plant diversity around the world. In the years ahead, they're also expected to play a vital role in our adjustment to another global trend.

I visited Fort Collins in the middle of May, but it could have been August. The thermometer hovered around 90°F (32°C), setting

FIGURE 7.2. This experiment on the International Space Station exposed 3 million basil seeds to the cold vacuum of space for more than a year. Later dispersed to scientists and school groups, the seeds sprouted successfully. PHOTO NASA MISSE 3, COURTESY OF NASA.

a string of daily records 20 degrees above the average. Two weeks earlier, another weather record had been set—for snowfall. In that context, my conversation with Chris Walters naturally turned to climate change. "It's already affecting how we collect and what we collect," she told me. I asked for an example, and she replied in a flash: "Sorghum. It's going to be huge." She explained how this tall, African grass was naturally adapted to a warm climate. "It's the hot, dry grain, and we'll all be growing more and more of it." Planning for that future, the seed bank's collection already contains 40,000 different sorghum samples.

If Chris is right, then seed banks will play a key role in the era of climate change, easing our transition to alternative, warm-weather crops. But they also protect agriculture against catastrophic events—wars, natural disasters, or political upheavals that can bring whole farming systems to a halt. In 2008, scientists unveiled a new

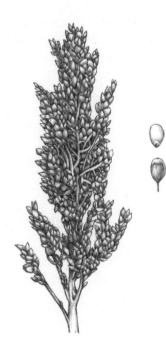

FIGURE 7.3. Sorghum (*Sorghum bicolor*). A hot-country grain native to Ethiopia, sorghum is expected to become increasingly important as the world adjusts to climate change. The kernels can be ground into flour, fermented to make beer, and even puffed as an alternative to popcorn. ILLUSTRATION © 2014 BY SUZANNE OLIVE.

international seed repository in the Norwegian Arctic. Carved deep into a mountainside in the Svalbard archipelago, it preserves seeds in cold, dry darkness with little need for additional refrigeration or other support from above. "If there are any big problems on the outside," its founding director noted, "this is going to survive." Dubbed the "Doomsday Vault," its opening made headlines around the world.

"Fear sells," Chris quipped when I mentioned the Svalbard project. But she quickly added that everyone in the seed community was grateful for the publicity. The attention raised the profile of their work and provided a needed boost in the constant struggle for funding. And running a seed bank is anything but cheap. While words like "vault" and "bank" imply simply turning the key and walking away, managing a seed collection requires constant activity. Even in cold storage, the samples steadily degrade and must be checked continuously to make sure they're still viable. "The original plan was every seven years, but we don't have the budget for that," Chris told me when we toured the germination lab. We stopped by a bench where a technician showed us trays of bean seedlings, each sprout carefully wrapped in damp paper towel. "So now we're on a ten-year cycle . . . but we don't have the budget for that either!"

Without regular germination tests, the seeds in any given sample could wink out before anyone noticed. "They die from an accumulation of insults," Chris explained. Small problems add up over time, like the aches and pains that everyone starts to feel as they age. Taken separately, none of these is serious, but when seeds pass a certain threshold their viability suddenly drops off to nothing. The trick lies in catching a sample before that happens, so that the seeds can be planted, grown to maturity, and then harvested to restock the collection. Regenerating older samples can keep a seed collection viable in perpetuity, but with varieties ranging from tropical cashews to winter-hardy kales, no single facility can handle all that planting.

"We don't do that part here," Chris said, sounding relieved. Instead, she and her team partner with over twenty regional seed banks and research stations in locations (and climates) as diverse as North Dakota, Texas, California, Hawaii, and Puerto Rico. They also collaborate with the seed vault at Svalbard and with an impressive facility for wild species managed by Kew Gardens. In fact, the number of seed banks worldwide is growing rapidly as governments, universities, and private groups recognize the threats posed by declining crop diversity and the loss of native plants. "There are over a thousand of us now," Chris announced toward the end of our day together. "It's becoming a movement!" Like any movement, seed banking has its villains and heroes. The villains tend to be faceless—large-scale patterns of habitat loss or trends in global agriculture. But in one case the role of "seed enemy" was played by a very recognizable historical figure: Joseph Stalin. Because when Stalin turned against the scientific community and began jailing Soviet scholars and intellectuals, his victims included the movement's first and most enduring hero, a brilliant botanist whose work influenced crop breeding for generations and paved the way for every seed bank that followed.

Though he is little known outside botanical circles, many regard Nikolai Vavilov as one of the greatest scientists of the twentieth century. The son of a wealthy industrialist, he survived the Bolshevik Revolution by virtue of his expertise. V. I. Lenin may have deplored the educated "intelligentsia," but he also believed in a science-based approach to modernizing Soviet agriculture. During the crippling grain shortages of 1920, Lenin diverted scarce funds from relief efforts to found the Institute of Applied Botany. "The famine to prevent is the next one," he famously told a colleague, "and the time to begin is now."

As the institute's first director, Vavilov received generous support for his plant breeding research and, by extension, his passion for seeds. He traveled widely and gathered samples by the ton, gaining a deep appreciation for how crops such as wheat, barley, corn, and beans varied from place to place—maturing early or late, surviving frosts, or resisting pests and disease. Better than anyone else in his generation, Vavilov understood how these traits could be stored indefinitely, in the form of seeds, and used to breed new varieties. He dreamed of developing crops specifically tailored to Russia's harsh climate, varieties that would end his country's persistent and deadly food crises. Within a few years, he transformed a tsarist palace in downtown Leningrad into the world's largest seed bank and research facility, supported by a staff of hundreds working in field stations across the country.

Unfortunately, Stalin did not share his predecessor's enthusiasm for scientific crop breeding, and he showed little patience for Vavilov's time-consuming methods. Soon after Lenin's death, the seed-bank program—and the Mendelian genetics on which it was based—fell out of favor. When another famine struck the country in 1932, Stalin threw his support behind the "barefoot scientists"—a cadre of untrained proletariat agriculturalists who promised quicker results. Vavilov found his research increasingly thwarted, and he was eventually arrested on trumped-up charges of sabotaging Soviet agriculture. He continued to write about seeds and crop plants in prison until his strength finally failed him. Neglected by his jailors, this champion of feeding the hungry suffered a final irony: he died of starvation.

But while Vavilov languished in prison, his ideas took on a life of their own. Soon seed banks based on the Russian model began springing up around the world. The United States broke ground in Fort Collins at the height of the Cold War, after the *Sputnik* launch inspired a widespread effort to "catch up" with Soviet science. Nazi Germany pursued a more direct route. During the siege of Leningrad, Hitler dispatched a special commando unit with instructions to secure Vavilov's seed bank at all costs and bring the collection home to Berlin. The city never fell, but the seed bank still faced a constant threat of looting by the starving populace. At least four devoted workers died from hunger without ever touching the thousands of packets of rice, corn, wheat, and other precious grains in their care.

Surprising stories of seed heroism continue to the present day. As US troops advanced on Baghdad in 2003, Iraqi botanists frantically packed samples of their most important seeds and shipped them to a facility in Aleppo, Syria. Everything that stayed behind was destroyed. Ten years later, the Syrians did the same thing, evacuating their entire collection mere days before Aleppo became a battleground in their own burgeoning war. Unfortunately, no amount of courage can save some collections. Somalia lost its two seed banks during the 1990s; Sandinista rebels looted Nicaragua's national collection; and invaluable strains of wheat, barley, and sorghum disappeared from Ethiopia's seed bank during the 1974 war that toppled Haile Selassie.

In light of this history, the high security and Cadillac-proof walls at Fort Collins start to make more sense. But while few people would argue that seeds aren't worth protecting, I hadn't heard Chris Walters or anyone else mention a fundamental irony underlying the whole seed-bank movement. Until very recently, crop diversity pretty much took care of itself, maintained by the same farmers, gardeners, and plant tinkerers that developed it in the first place. Wherever people farmed, they bred local varieties and kept them "banked" in their fields, replanting and refining them season after season. Saving that diversity only became an issue after the advent of industrial agriculture, with its focus on high yields from a few varieties grown on a massive scale. As impressive and necessary as seed banks have become, they are in many ways an elaborate fix to a problem of our own making.

"I agree completely," Chris said when I posed this dilemma. "The best kind of conservation is in situ." For crops, that means in a farmer's field; for wild species, it means in a healthy expanse of natural habitat. "But that's not always possible," she went on simply, showing the pragmatism that makes her such a good scientist. "Seed banking is something we *can* do, and so we should. It's a way of buying time."

Because of dormancy, boosted by refrigeration, seed banks can indeed buy a great deal of time. But while they will always be a vital resource for plant research and breeding, there is still the question of what they're buying time for—what changes in human activity would lead to the kind of in situ conservation that Chris was talking about? Part of the answer lies not in a laboratory or a cryogenic tank, but on a small farm outside the town of Decorah, Iowa, population 8,121. There, for nearly forty years, a group of dedicated gardeners have kept thousands of different vegetable varieties growing, not just in their own fields, but in garden plots around the world.

"Our collection is a living collection," Diane Ott Whealy told me. "Heirloom vegetables aren't like heirloom furniture or jewelry—you can't just take them out once in a while and dust them off. The best way to preserve these seeds is to plant them."

I reached Whealy at her office on the farm, an audibly busy place where people interrupted our conversation regularly to ask questions or schedule meetings. Like Fort Collins, the facility at Decorah boasts climate-controlled rooms generously stocked with seeds. But unlike the government establishment, Whealy's group also runs an 890-acre farm, operates a mail-order seed business, and coordinates a growing global network of "backyard preservationists." If Chris Walters can call 1,000 seed banks a movement, then the 13,000 members of the Seed Savers Exchange should count as a revolution. "We're a people's seed bank," Diane said simply, "dedicated to identifying, preserving, and distributing heirloom vegetables." But while she and her colleagues do maintain a traditional collection (with duplicate samples at Fort Collins and Svalbard), their overarching goal is to reconnect seeds with people, helping gardeners and farmers collect, trade, and, most importantly, *plant* heirloom seeds, year after year.

Diane and her then-husband, Kent Whealy, founded Seed Savers in 1975, inspired in part by the seeds of an unusual purple morning glory she inherited from her grandfather. ("That morning glory has a lot of personality," she told me. "Just like grandpa.") From a card table in their living room, the project quickly grew into a world-wide network of passionate seed collectors. "There's a great emotional attachment to seeds," she explained. "When people started sending us samples, they often included a recipe. Yes, they wanted their varieties preserved, but they also wanted them to be grown, harvested, eaten—celebrated as food!" From the beginning, people also joined the exchange to meet other seed savers. An annual picnic evolved into a three-day seed conference and festival, and the exchange's first seventeen-page newsletter grew into a tome the size of a phonebook listing more than 6,000 varieties for sale or trade, many of them available nowhere else.

From a biological perspective, Seed Savers provides a vital complement to the effort at Fort Collins. The larger facility holds a vast diversity, but it's one that rarely changes—the seeds are only grown when the staff needs to restock the shelves. "Keeping seeds planted allows those varieties to continue adapting," Diane explained. "Even without climate change, plants need to adjust to local conditions." By virtue of their constant gardening, the seed savers do more than maintain garden diversity. They're allowing the plants to evolve, helping create new variation that will stock the gardens and seed banks of the future.

At the end of our conversation, I asked Diane if she could envision a time when the work would be through, when enough people would be planting enough varieties to make seed banks unnecessary. "No, it's never done," she said, and laughed with the ease of someone who has found her calling. "We'll be seed pushers forever."

Part of the success of the Seed Savers Exchange lies in the willingness—even eagerness—of its membership. Any gardener, or anyone who has lived with a gardener, knows that planting and harvest are only part of the process. In our household, one of the most exciting gardening moments of the year comes in the dead of winter, with the arrival of the seed catalogs (including the hefty Seed Savers Yearbook). For Eliza, this marks the official start of a new season. While cold rain and windstorms rage outside, she pages contentedly through thousands of different vegetable and flower varieties, choosing the next year's crops. Noah loves these catalogs, too, and it's not unusual to find a few well-thumbed copies mixed in with *Goodnight Moon*, *Make Way for Ducklings*, and the other classics tucked beside his bed.

Though fascinated by anything to do with seeds, I consider myself less a gardener than a garden "enabler." For Eliza (and now Noah), gardening is both passion and pleasure, a fruitful addiction that I'm happy to support. If I focus on splitting firewood, cutting grass, and other household chores, it frees up more time for them to spend in our ever-expanding garden. And since we all share in the harvest of delicious fruits, vegetables, and berries, the arrangement works quite nicely. There is one patch of ground, however, that I help cultivate every year.

Like Eliza, my mother had a passion for gardening, and like me, my father always played a greater role in eating the produce than he did in the watering and weeding. But since Mom died, Noah and I have visited my dad every springtime to help him replant her garden, at least in part. Dad and I take solace in tilling and sowing the same soil she once worked, and in Noah's unbridled enthusiasm for the whole affair. It's a ritual of remembrance enriched by the curious biology of seeds—by dormancy, and the desire to coax life from something that appears so lifeless. That abiding mystery often brings even the most serious discussions of seed science to a place where fact meets philosophy.

Before leaving Fort Collins, I asked Chris once again to help me understand the metabolism of a dormant seed. Carol Baskin had told me that the cells were still active, but at a very reduced level. Chris held a different view. Dormant seeds do change over time, she admitted, but it wasn't necessarily a sign of cell activity in the traditional sense. "I think what we're seeing is just the natural breakdown of organic compounds," she said, her years of chemistry coming to the fore. "It's like an expiration date on a prescription medicine. The chemicals in the drug simply degrade until they stop working. Seeds are the same way."

I knew Chris was speaking from experience. She had an entire research program devoted to measuring the air around seeds, documenting changes in the chemical signatures they give off as they age. But it still bothered me. How could seeds be alive without any discernable metabolic activity?

"I'll answer that question with a question," she said immediately. "Does metabolism define life? If seeds are alive but aren't metabolizing, then maybe we need to rethink our definition of what it means to be alive."

After decades of study and thousands of years of planting and harvest, seeds retain the ability to challenge our most basic ideas. That makes them fascinating not only as a research topic, but also as a metaphor for life and renewal. It's no coincidence that "seed" appears in more than three hundred English words and phrases, from the obvious *seed-corn* (grain saved for planting) to the less intuitive *hag-seeds* (the children of a witch). In fact, you could say that Chris had left me with a *thought-seed*, the kernel of a notion that may yet sprout, blossom, and bear fruit. I'm still thinking about what she said, because the only way to really know if a seed is alive, even at the National Seed Bank, is to plant it and see if it grows.

While people may speculate about the life contained in seeds, the flowers, shrubs, herbs, and trees that produce them have no room for doubt. Their faith is evolutionary and absolute. Nothing shows that better than the topic we'll turn to next, the incredible (and incredibly useful) ways that plants defend their seeds. That spark of dormant life may be hidden and hard to measure, but mother plants will do almost anything to protect it.

Seeds Defend

Never come between a lioness and her cubs.

—Traditional proverb

CHAPTER EIGHT

By Tooth, Beak, and Gnaw

Oh rats, rejoice!
The world is grown to one vast drysaltery!
So munch on, crunch on, take your nuncheon,
Breakfast, supper, dinner, luncheon!

— Robert Browning,
The Pied Piper of Hamelin (1842)

Appendix F of the International Building Code stipulates requirements for keeping rats and other rodents out of all habitable dwellings. These include two-inch (five-centimeter) slab foundations, steel kick-plates, and tempered wire or sheet-metal grating over any ground-level opening. Conditions for grain storage or industrial facilities can be even stricter, involving thicker concrete, more metal, and curtain walls buried two feet below grade. In spite of all this, rats and their relations still consume or contaminate between 5 and 25 percent of the world's grain harvest, and regularly gnaw their way into important structures of all kinds. In 2013, a trespassing rodent shorted out the switchboard at Japan's ill-fated Fukushima nuclear plant, sending temperatures in three cooling tanks soaring and nearly setting off a repeat of the 2011 meltdown. The story made headlines around the world, with journalists, bloggers, and TV commentators all wondering what makes rats so interested in electrical wires. But the real question isn't about what rodents like to eat; it's about how difficult it is to stop them. Why on earth should a rat be able to chew through concrete walls in the first place?

The name "rodent" comes from the Latin verb *rodere*, "to gnaw," a reference both to the way rodents chew and to the massive incisors that help them do it so well. These teeth evolved in small mouse- or squirrel-like creatures approximately 60 million years ago. That's approximately 60 million years *before* the invention of concrete, Plexiglas, sheet metal, or any of the other manmade materials that rats and mice now chew through. Experts still argue about the exact origin of rodents, but there is little doubt about what those big teeth were good for. While the family tree now includes oddballs like beavers, who chew wood, and naked mole rats, who use their teeth for digging, the vast majority of rodents still make much of their living the old-fashioned way: by gnawing seeds.

Before rodents came along, the ancestors of trees like oaks, chestnuts, and walnuts got by with little winged pips that offered scant protection from chewing. Fossils of these seeds look like lumpy flecks of chaff, insubstantial wisps designed to flutter a bit as they fell. Once the gnawing began, however, these plants and their rodent predators entered a virtual arms race: stronger teeth led to harder seed coats and vice versa, changing those ancient seeds into the acorns and thick-shelled nuts we're familiar with today. (Other seeds responded by getting even smaller, in the hopes of being swallowed whole, or ignored altogether.) For the trees, rodents posed an evolutionary dilemma: the chance to get their seeds dispersed balanced against the risk of losing them entirely. For rodents, unlocking the nutrition in seeds turned out to be an evolutionary gold mine: they quickly became the most numerous and diverse group of mammals on the planet.

The notion of coevolution implies that change in one organism can lead to change in another—if antelope start running faster, then cheetahs must run faster still to catch them. Traditional definitions describe the process as a tango between familiar partners, where each step is met by an equal and elegant counter-step. In reality, the dance floor of evolution is usually a lot more crowded. Relationships like those between rodents and seeds develop in the midst of something more like a square dance, with couples constantly switching partners in a whir of spins, promenades, and do-si-dos. The end result may appear like quid pro quo, but chances are a lot of other dancers influenced the outcome—leading, following, and stepping on toes along the way. No one knows the exact sequence of events that gave us strong-jawed rodents and thick-shelled seeds; the story played out long ago and left only general clues in the fossil record. But few experts believe their sudden and simultaneous rise was mere coincidence.

In many cases, the relationships that developed became mutually beneficial—the gnawers got something to eat and dispersed a few of the plant's seeds in the process. Hunger alone drives the rodent side of this equation, but for plants it's like walking a tightrope. Their seeds must be attractive enough to be desired, but tough enough so that they can't be devoured on the spot. A hard shell forces rodents to carry seeds away and gnaw them open later, in the safety of a burrow. Ideally, the rodents then forget where they've hidden things, or perish before they get around to eating them. Take the example of Beatrix Potter's book *The Tale of Squirrel Nutkin*. Scholars think she wrote it as a commentary on Britain's class system, but it's also a story about seeds: if the squirrels on Owl Island gather and stash away nuts, and if Old Brown the owl attacks the occasional squirrel, then some of those nuts will go uneaten and the next generation of oaks and hazels will live on. (Nutkin managed to escape with only the loss of his tail, but we must assume that Old Brown is more successful on other attempts.)

Potter set her story in England's Lake District, but if she had lived in Central America she would have put it right where I did my

FIGURE 8.1. The busy squirrels from *The Tale of Squirrel Nutkin* (1903), Beatrix Potter's classic story about gathering (and dispersing) the acorns and hazelnuts of Owl Island.

doctoral research, under the spreading boughs of an *almendro* tree. There, little Nutkin would have found not only plenty of squirrels to keep him company, but also other rodents: pocket mice, rice rats, climbing rats, and spiny rats, as well as pacas and agoutis, which look more or less like guinea pigs the size of small dogs. Like me, all of these species came to *almendro* trees in search of seeds. Unlike me, the rodents had been at it for thousands, if not millions, of years. (A dissertation only *feels* like it takes that long.) With so many gnawing creatures hanging around, it's no wonder *almendro* developed a shell hard enough to challenge a graduate student. But the nuances of seed defense rarely stop at physical protection alone. The ecology of this one rainforest tree makes it clear why so many seeds are stony, and why it takes a lot more than concrete to stop a hungry rat.

An *almendro* seed measures two inches (five centimeters) long and slightly more than an inch (two and one-half centimeters) wide, with smooth sides and tapered ends that give it the appearance of a giant throat lozenge. Like the pit of a peach or plum, this seed includes an extra layer of stony shell, with the soft nut tucked safely inside. The surrounding flesh of the fruit is thin and brownish-green, but sweet enough to attract a wide array of monkeys, birds, and bats. At the height of the season, dozens of species gather around *almendros*, foraging in the canopy and feasting on the bounty that drops to the ground below. But among all these fruit-eaters, only one large bat carries its meal away from the tree. So if an *almendro* wants its young dispersed, it must also concentrate on the creatures that eat its seeds. And while it may be hard to think of trees as intelligent (at least outside of J. R. R. Tolkien stories), the system *almendros* have developed seems careful, calculating, and nearly perfect.

From a plant's perspective, not all potential dispersers are created equal. When I collected *almendro* seeds, for example, I carted off large quantities and traveled great distances, but then systematically destroyed every one of them for my research. Even if I'd planned on sprouting the seeds, my laboratory was at a university in northern Idaho, hardly the right habitat for rainforest trees. At the other end of the spectrum, smaller rodents, like rice rats and pocket mice, lack the strength to move *almendro* seeds more than a foot or two. Invite them to the feast and the tree's progeny would die without ever leaving home. Excluding small, ineffective seed predators and limiting the damage from large ones requires a shell with just the right level of defenses, one that optimizes what ecologists call *handling time*.

For *almendro*, the ideal shell turns out to be a woody husk that measures over a quarter of an inch (seven millimeters) thick at its

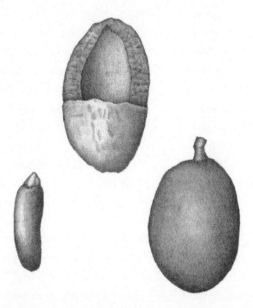

FIGURE 8.2. *Almendro* (*Dipteryx panamensis*). Seeds of the mighty *almendro* tree lie within one of the toughest shells in nature, a defense against the gnawing teeth of rodents. The shell is pictured at the top, partially cut away in cross section. An extracted seed is shown on the left, with a whole fruit on the right. ILLUSTRATION © 2014 BY SUZANNE OLIVE.

widest point, twice the heft of a plum pit or a peach stone. The walls include additional protections: a layer of resinous crystals, much like the ground glass that exterminators add to concrete when they want to plug a rat hole. But in this case the seed isn't trying to prevent gnawing entirely, just slow it down. For the average squirrel, chewing through the crystal-filled husk of an *almendro* takes at least eight minutes, and sometimes as much as half an hour. That's a huge time investment for an animal that needs to locate and eat between 10 and 25 percent of its bodyweight every day just to survive. An *almendro* seed is worth the effort, but just barely. Spiny rats and smaller rodents rarely bother—not necessarily because they can't, but because it's not worth their while. The challenge and time involved would exhaust them to a degree that not even the reward of a large nut could repay. In this context, the strength and thickness of *almendro* shells seem perfectly adapted to reserving those nuts for squirrels, agoutis, and pacas—the large rodents most capable of carrying them away. Making them actually do so, however, lies beyond the control of the tree. That incentive must come from other players in the dance.

Once I perfected my mallet-and-chisel technique, I learned to cleave an *almendro* seed and neatly extract the nutmeat in less than a minute. That put me well ahead of squirrels, but it wouldn't have seemed nearly fast enough if I'd been opening seeds in a dangerous setting—a crocodile pit, for example, or a pen full of hungry wolves. That's the dilemma faced by rodents. Because as surely as an *almendro* tree attracts seed-eaters, it attracts the *eaters* of seed-eaters. I knew from experience that fer-de-lance hang around *almendro* trees, and so do other rodent-loving snakes like bushmasters and boa constrictors. I once watched a Semiplumbeous Hawk carry off something small and furred in broad daylight. If I'd stuck around until dark, I might have seen half a dozen different owls, as well as ocelots, margays, and jaguarundis, all of them attracted by the concentration of tasty prey. A friend of mine studying mammal communities once showed me a stack of photographs taken by remote cameras scattered throughout the forest—flash snapshots of surprised-looking jaguars, pumas, big weasels, and others. There were even a few hunters and their dogs. He asked me if anything seemed familiar, and then I saw it: time and again the backdrop included an *almendro* trunk, and the ground was littered with seeds. In the rainforests of Central America, the community of animals drawn to a good *almendro* crop doesn't stop with fruit-eaters, seed-eaters, and predators. It includes all of the people who seek them: scientists, hunters, birdwatchers, and anyone else looking for a piece of the action.

In the face of all this commotion, much of it fanged and hungry, squirrels and other rodents usually treat *almendro* trees like a drive-through. They pick up a meal and carry it away before stopping to eat—forty feet (twelve meters), fifty feet (fifteen meters), and sometimes much farther. Agoutis, in particular, turn out to be vital dispersers. They not only move seeds a long way, they bury them for safekeeping in tidy little holes throughout their home range, a habit with the pleasing name of *scatter-hoarding*. From the tree's perspective, this fits the bill nicely: a creature that moves seeds and plants them, and then stands a good chance of being killed off by one of the many predators lurking nearby.

This pattern repeats itself with different rodent and plant species all around the world, providing ample evolutionary incentive for nut-like seeds to develop their thick, hard shells. In fact, any trait that adds to handling time can be an advantage, which is probably why walnuts have that brainy, convoluted shape that's so irritatingly difficult to remove in one piece. Rodents, too, have responded with more than just strong teeth, developing bulging, high-capacity cheek pouches to carry off numerous seeds at once, as well as an uncanny ability to sniff out and discard diseased or worm-infested nuts before bothering to gnaw them. Like so many evolutionary stories, the impact of rodents on seed defense is more than a bilateral arms race. It involves a whole suite of relationships and species, with give-and-take on all sides. For the *almendro*, my research showed not only how elaborate that system can be, but how quickly it can fall apart.

In a healthy rainforest, crossing the muddy ground near large *almendro* trees feels like walking on lumpy gravel—gnawed, split, and otherwise discarded shells carpet the ground underfoot. I counted them by the thousands and rarely found an intact seed, let alone a young tree. With such a concentration of gnawers around, the only seeds that sprouted and grew into saplings were those that got themselves dispersed far away. In patchy forest, however, hunting and other disturbances took a heavy toll on large rodents, and there was hardly any sign of gnawing or scatter-hoarding. Seeds simply germinated where they fell, leaving every adult tree ringed by a thicket of its own children. In the short term, this scenario meant bad news for the next generation—baby trees don't fare well in the shade of their parents. From an evolutionary standpoint, it put *almendro* in a bind—removing its partners in the dance left it with a seed too hard for anyone left in the forest to chew.

Studying *almendro* taught me that plants defend their seeds with a complex calculus where protection is only one of the variables. But it also left an obvious question unanswered: Just how hard is an *almendro* shell? Harder than concrete? I found the answer while writing this chapter.

Though I finished my dissertation years ago, a person doesn't invest that kind of time in a project without taking home a few souvenirs. Dried now to a rough honey-brown, the *almendro* shell I keep on my desk still bears the telltale grooves of rodent bites at one end. To test it against concrete, I simply stepped outside of my office and crawled under the porch. The Raccoon Shack rests on a foundation of concrete pier blocks, the standard variety with a built-in bracket available at any home supply store. I placed my *almendro* shell edgewise against a pier block—like a chisel—and gave the thing a good solid whack with a hammer. It didn't surprise me in the least to see cracks form in the concrete: if rodents evolved to gnaw, and *almendro* responded with one of nature's hardest seeds, then *almendro* shells should be very nearly as tough as rat teeth. A few more whacks produced a sizable concrete chip that flew loose and landed in the soil below. I reached down to pick it up, careful to avoid the less savory items lying around under the porch: droppings and tattered feathers from our chickens, and half a dozen empty rat traps. One look at those traps annoyed me, and I made a mental note to return that evening and re-bait them with nut butter.

"People won't believe you," Eliza warned, smiling when I told her what was happening under the Raccoon Shack. But as Oscar Wilde once observed, "life imitates art far more than art imitates life." And the fact remains that while I was sitting at my desk, writing about rodent teeth and seeds, a version of that same drama was playing out directly beneath my feet. Attracted by the grain in our nearby chicken coop, an extended family of Norway rats had moved into the crawlspace under the Raccoon Shack. They gained entry by chewing a neat hole through a sheet of 23-gauge galvanized steel hardware cloth. Once inside, the rats had a cozy home base from which to stage raids on any nearby edibles. They soon discovered my pea bed, where I'd stupidly left the entire harvest for my Mendel experiment drying on the vine. By the time I figured out what was going on, the rats had decimated my Bill Jump Soup Peas and put a sizable dent in the Württembergische Wintererbse. The sad remnants that Noah and I picked through totaled a scant three cups, but luckily they included enough successful crosses to continue the experiment for another (better-protected) season.

Losing my peas to the rats turned out to be a valuable lesson—as Mr. Wilde also noted, "experience is the name we give our mistakes." First of all, I gained another insight into the methods of that cagey old monk. Unless the Monastery of St. Thomas maintained an army of cats, Mendel must have built a safe place to dry his harvest. It wouldn't surprise me if his lost journals and papers contained detailed plans for a rat-proof granary. More importantly, I learned that even in the artificial setting of my pea bed, with a domestic vegetable and a nonnative rodent, the same rules apply. When the rats sniffed out my vines, they did their job perfectly, using the precise logic found in any rodent/seed interaction. Bill Jump peas mature slowly and hadn't fully dried yet, making them comparatively easy to chew. Those were eaten on the spot. When I tried biting into a winter pea, however, I nearly broke a molar. They required more handling time, and, according to theory, should have been hauled away for safe gnawing. Sure enough, when I opened up the crawlspace under the building, I found a huge pile of empty pods and winter-pea seed coats. (The opposite of a scatter-hoarder, the Norway rat stores all its seeds in one place and is known in biology circles as a "larder-hoarder.")

In the weeks I spent baiting traps below the Raccoon Shack, I found myself wishing that rats had never evolved. But even in a

FIGURE 8.3. The larderhoard of a Norway rat family underneath the Raccoon Shack, final resting place for most of the pea harvest from my Mendel experiment. PHOTO © 2013 BY THOR HANSON.

world without rodents, something probably would have been after my peas. Once mother plants began packing lunches for their babies, everything from dinosaurs to fungi wanted a taste, and the evolution of seed defenses became inevitable. The relationships sometimes find a balance, but not always. *Almendro* trees appear to have the rodent situation figured out, but they must not have planned for peccaries, aggressive wild pigs whose massive molars can split and crush the seeds with ease. Even worse, the Great Green Macaw specializes on *almendro*, nesting in the trees and gorging on the seeds, which it can handily crack with a bill specifically adapted to the purpose. Among seed predators, birds have one of the longest evolutionary histories. They descend from dinosaurs, some of whom developed seed-crushing organs over 160 million years ago. Paleontologists know this from fossils containing telltale clusters of *gastroliths*, the

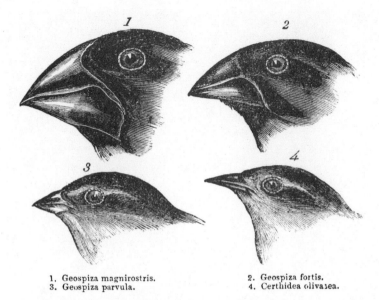

1. Geospiza magnirostris.   2. Geospiza fortis.
3. Geospiza parvula.   4. Certhidea olivasea.

FIGURE 8.4. This classic illustration by John Gould shows some of the diversity of beak shapes in Darwin's Galapagos finches. Charles Darwin, *Journal of the Beagle* (1839). WIKIMEDIA COMMONS.

distinctive little stones found inside gizzards. Modern birds still depend on grit to grind their food, and the strongest gizzards occur in seed-eaters—everything from chickens to canaries, grosbeaks, jays, and perhaps the most famous group of birds in the world.

For Charles Darwin, the finches of the Galapagos Islands appeared like a gaggle of unrelated species, more notable for their tameness than anything else. As he recorded in the field, "Little birds . . . will alight on your person & drink water out of a basin held in your hand." It wasn't until his specimens reached ornithologist John Gould, who had worked on parrots and was very familiar with seed-cracking bills, that their close affinity came to light. As famously told in Jonathan Weiner's *Beak of the Finch*, biologists have since learned that seasonal changes in seed abundance produce measurable evolutionary changes in the finches. Differences of less than half a millimeter in bill length determine which birds can crack the toughest seeds and which can't. In times of scarcity, that distinction is a matter of life and death, and as a result, the bills of whole populations can change *in a single generation*. The fact that natural selection can play out so quickly helps to explain how one original Galapagos finch could have morphed into thirteen species, some with seed-crushing bills, some that sip nectar, and some that eat fruit or insects. There are also cactus-flower probers, and a finch whose beak can hammer bark like a woodpecker. Played out globally, the Galapagos scenario helps the mind grasp how specializing on seeds (or other foods) can have such evolutionary influence. According to one theory, overcoming the physical challenge of eating hard-shelled seeds may have even produced the distinctive shape of the human skull.

As a child, I subjected my skull to a typical variety of sports. Though I eventually settled on swimming, I passed several seasons playing soccer and baseball, and even briefly threw my small frame into the melee of American football. One similarity among all these activities was the healthful snack served during practices and games: fresh oranges sliced into wedges. And, given that snack, we young athletes would immediately stuff wedges into our mouths, skin side out, and run around hooting like chimpanzees. Try this now and you'll find that it does create an undeniably ape-like impression. But it's not the wide orange smile that does it. I spent two years studying mountain gorillas in Uganda, and while they expressed themselves in all sorts of ways, I rarely saw them leer. The orange-wedge trick works because it reshapes the skull, giving a forward, snout-like projection to the jawbone. All other apes, and most ancient hominids, share that structure. But in human ancestors, the face began to flatten, and that's where the seeds come in.

"There was a radical shift around 4 million years ago," explained David Strait, a professor of anthropology at the State University of New York. Modern human faces appear flat because our bones are small, he told me, probably an adaptation for eating soft, cooked foods. But it was another dietary shift that started the ball rolling. "The facial reinforcements," he said, "the large cheekbones and muscle attachments, the size and shape of the teeth—all point toward producing and withstanding high loads." Just the kind of "high loads" that come from cracking the shells of hard seeds and nuts.

For much of the past decade, Strait and his team have argued that habitually biting large, hard objects like nuts explains the changes in ancient skulls. Their computer models show the digitized facial bones of *Australopithecus*—an extinct hominid best known for the "Lucy" specimen—happily munching away, with the force of each bite distributed particularly above certain teeth. It's a habit we maintain. To revisit the sports analogy, spectators at a game don't eat orange wedges. Instead, they litter the stands with hot dog wrappers, drink cups, and, invariably, the empty shells of dry roasted peanuts. Next time you have a bag, notice which teeth you use to crack open the tough ones. Chances are you'll position that nut on the side of your mouth, right behind the canines, where your skull best absorbs the force of the bite. Those are the premolars, and if Strait is right, using them for nutshells is a deep, evolutionary instinct.

"A lot of my colleagues don't believe me," he laughed, "and that's fine!" Critics of Strait's "hard-food" theory point to chemical analyses and patterns of tooth wear that suggest a diet dominated by grasses or sedges. But Strait doesn't see this as a conflict. Ancient hominids might have gobbled up all sorts of things when food was abundant, but just like the Galapagos finches, what really mattered was getting through the tough times. "Nuts were the fallback food," he said, and fallback foods can drive evolution because the stakes are so high. "Soft foods and fruits are wonderful and sweet," he told me, with the ease of someone practiced at making his point, "but when those run out, you either have to move, eat something else, or die." In those terms, it makes perfect sense for the hominid face to reorganize itself around a nut-chomping, premolar bite.

If the habit of eating hard seeds did influence our skulls, much in the same way it shaped finch beaks and rodent jaws, then how might human chewing have impacted seeds? Toward the end of our conversation, Strait hinted at an answer. He mentioned new research showing how the micro-structure of seed shells mirrors that of tooth enamel. In each case, the cells lie in tight rows of ray-like rods and fibers, as if both sides arrived at the same engineering solution to resist the impact of the other. He also passed along a paper about a Southeast Asian seed so strong it can barely germinate—the halves of its dense husk cling together at the very limit of what a growing shoot can split apart. Yet in spite of this, the seed still falls prey to beetles, squirrels, and the occasional orangutan. It's a reminder that physical defenses can only go so far. From *almendros* to peanuts, the story is the same: no matter how strong a seed's shell, there will always be a rat, parrot, or sports fan nearby who evolves a way to break through. Which is, of course, why shells are only the tip of the iceberg. If plants could successfully protect their babies just by building a better box, then there would be no point in drinking coffee, Tabasco sauce would be tasteless, and Christopher Columbus would never have sailed for America.

CHAPTER NINE

The Riches of Taste

All hot! All hot!
Pepper pot! Pepper pot!
Makes backs strong,
Makes lives long,
All hot! Pepper pot!

—Traditional Philadelphia street-vendor cry

"You come from far away," the old man said, "where the devil left his jacket." His speckled grey pony shifted beneath him, shaking the braided circlets of blue, red, and green leather that hung from its homespun bridle. I tried to meet the man's eyes, but his stare never wavered from the air just above our heads. So I smiled at the horse, feeling foolish. They were blocking our path, but as visitors to a tribal area we needed permission to pass. The conversation didn't seem to be going well.

"We've been mistreated ever since you came here," he said, and I was confused. Hadn't we just arrived? I wasn't even sure if *almendro* grew in this forest. Then he clarified: "You and your Columbus."

In biology, scouting new research sites occasionally involves sneaking a look at a promising field or forest without leave. But those few words reminded me that I was trespassing on an entire continent.

128

Even the Costa Ricans with me didn't count as locals—their ancestors came from Spain, following the trail blazed by Columbus himself when he dropped anchor at nearby Puerto Limón in 1502. Eventually, the old man nudged his pony to the side of the road and, having made his point, graciously gave us welcome. We didn't find *almendro* that day and I never returned, but his words stayed with me. Centuries may have passed, but people still journey to the ends of the earth searching and seeking. And later I realized that Christopher Columbus and I did have one thing very much in common: we'd both come looking for seeds.

"Simply standing on the beaches," the great explorer once wrote, " . . . we find such traces and clues of spices that we have reason to believe much more will be found in time." The logbook from his first voyage contains no fewer than 250 botanical descriptions, often-detailed accounts of the crops, trees, fruits, and flowers he encountered in the Caribbean. But while the list included plants (and seeds) that would redefine European cuisine and commerce, from corn to peanuts to tobacco, Columbus showed hints of disappointment as those first weeks wore on. "I am sorry to say I do not recognize them," he wrote, after inspecting the herbs and shrubs of Isabella Island (now Crooked Island). Several days later, the flora made him "extremely sorry," and in another passage he laments a forest of fragrant but unfamiliar trees: "It grieves me extremely that I cannot identify them." Columbus felt worried, because while his ships may have bumped into a New World along the way, the admiral had promised his sponsors a much different outcome. Queen Isabella, King Ferdinand, and all of his other noble backers expected more than tales of discovery—they wanted to get rich. They had invested in a new trade route to Asia and expected a return paid in Asian products: gold, pearls, silk, and, above all else, the exotic spices that grew nowhere else. Unfortunately, neither Columbus nor anyone traveling with him had any idea what they looked like.

In the fifteenth century, spices reached Europe only after passing through so many intermediaries, along such a complex network of

FIGURE 9.1. Christopher Columbus scoured the New World for any sign of Asian spices, filling the logbook from his first voyage with more than 250 botanical descriptions. While nutmeg, mace, and black pepper eluded him, he did bring home the tasty seeds of allspice and chili peppers. COLUMBUS TAKING POSSESSION OF THE NEW COUNTRY, L. PRANG & COMPANY, 1893. LIBRARY OF CONGRESS.

Asian and Arab trade routes, that people on the receiving end saw only the finished products, with little hint of how or where they grew. Popular myths suggested looking for flaming trees guarded by serpents, examining the sticks in Arabian bird nests, or harvesting twigs and berries from Paradise itself. Marco Polo, at least, attributed spices to real plants growing in real places: India and the Moluccas. But those were little more than names in a story to people back home—people didn't understand their geography, let alone their flora. Whenever Christopher Columbus encountered a new plant, he must have sniffed its bark for hints of cinnamon, tasted the flower buds hoping for cloves, and scratched at a root in search of ginger.

Then he would have turned his attention to its seeds, home to the most valuable spices of the day—nutmeg, mace, and pepper.*

Scholars often compare the historical craving for spices to the modern appetite for petroleum. Both situations combined limited supply with virtually limitless demand, creating commodities that anchored the global economy. But where oil reserves already show signs of dwindling, the harvest of spices remained constant and even increased, extending their reign for centuries. Tracing that story reads like a history of commerce, exploration, and civilization itself. In ancient Egypt, for example, peppercorns from India's Malabar Coast somehow made their way up the nostrils of dead pharaohs— they were the royal embalmers' most prized preservative. When Rome found itself surrounded by Visigoths in AD 408, the barbarians demanded 3,000 pounds of pepper as part of their ransom to end the siege. King Charlemagne issued a decree in 795 that called for cumin, caraway, coriander, mustard, and a host of other flavorful seeds to be grown in gardens throughout the Frankish Empire. Paying feudal tithes with spices became common during the Middle Ages, and the practice still persists: when the current Duke of Cornwall (alternatively known as the Prince of Wales), England's Prince Charles, officially accepted his title in 1973, he was presented with a pound each of pepper and cumin.

Some of the most telling spice statistics of all, however, boil down to simple economics. In fact, they sound like the prospectus for a stock offering. During its first fifty years, the Dutch East India Company dominated world trade in nutmeg, mace, pepper, and cloves, experiencing one of the greatest eras of profit in the history of business. Gross margins rarely dipped below 70 percent, and the company

---

*Nutmeg and mace both come from a tree native to Indonesia. Nutmeg is the seed itself, while mace grows as a fleshy red seed appendage called the *aril*. Pepper comes from a rainforest vine native to the west coast of India. Black pepper includes the seed and a thin layer of shriveled fruit tissue; white pepper is the same thing with the fruit layer removed.

paid lavish dividends, both in cash and in spices. Original shareholders who held on to their stock enjoyed average annual returns above 27 percent for forty-six years, a rate that would have turned a modest $5,000 investment into a fortune of more than $2.5 billion over that time. (For comparison, Exxon Mobil—currently the world's most profitable enterprise—earns total returns of around 8 percent a year.) With that kind of money at stake, it's no wonder the Dutch happily gave Manhattan to the British in 1674 in exchange for a tiny, nutmeg-producing island in Indonesia. And it's not surprising, either, to learn that one of the only pirate chests ever recovered—buried by Captain William Kidd in 1699—contained not gold or silver, but a few bolts of fancy cloth and a bale of nutmeg and cloves.

In terms of exploration, however, nothing puts Christopher Columbus's spice worries in better context than the results of a voyage only slightly less famous than his own. When Ferdinand Magellan set sail a quarter century after Columbus, he promised his backers the same result: a direct western trade route to the Spice Islands. Three years later, four of his five ships were lost and Magellan was dead, along with his second-in-command, third-in-command, fourth-in-command, fifth-in-command, and over two hundred crewmembers. Yet when eighteen survivors limped into Seville on the sole remaining vessel in 1522, they had more to show for their trouble than a global circumnavigation. Their small cargo included nutmeg, mace, cloves, and cinnamon from the Malaccan island of Ternate. When sold, the spices brought in more than enough cash to pay for the lost ships and compensate the families of the deceased, turning the journey into one of discovery *and* profit. Without finding spices, Christopher Columbus would never manage that feat.

History remembers Columbus for his epoch-making first trip across the Atlantic and for helping usher in a new era of exploration and conquest. But people often gloss over the fact that he returned to the New World three more times, searching in vain for spices, gold, or other valuable commodities. On the second voyage, he found every member of his new colony on Hispaniola murdered by natives. He returned from the third journey in chains, accused of tyranny, and he ended the fourth expedition shipwrecked on Jamaica for more than a year. As one biographer put it, "money was continually being spent on ships and supplies; where was the return for it? . . . What about the Land of Spices? . . . To the most impartial eyes it began to appear as though Columbus were either an impostor or a fool." While others suspected he'd found something new, the admiral stuck to his claim that the Caribbean Islands and surrounding coastlines were indeed parts of Asia, and that spices— not to mention Japan, China, and India—would turn up in time. But though Columbus would go to his grave without knowing what continent he'd discovered, one thing is certain: he knew he'd found the wrong pepper.

"There is also much *aji*, which is their pepper and is worth more than our pepper," he wrote, after dining with the locals on Hispaniola. Though he'd never seen black pepper growing, the difference in flavor and pungency, not to mention the shape and color of the seeds and fruits, told him that this spice was different. His claim about its value can be chalked up as good old-fashioned spin-doctoring. In the waning days of that first voyage, he needed to put the best face on whatever seeds, plants, and scraps of gold he'd cobbled together for a cargo. But in retrospect, Columbus's words seem prophetic, because by most measures the chili peppers he brought back across the Atlantic have gone on to become the most popular spice in the world.

Dried and ground or added whole, the fruits and seeds of *Capsicum* chili peppers now flavor everything from Thai curries to Hungarian goulash to African groundnut stew. From four wild species native to the New World, over 2,000 cultivars have been developed, ranging in spiciness from the mildest paprika to the fieriest habañero and beyond. (Bell peppers also come from this stock, but are bred for size and sweetness instead of pungency.) One in four people around the world eat chili peppers daily, and in a twist that might have pleased the frustrated admiral, they've replaced black pepper as the hot spice of choice throughout India and Southeast Asia. He may have failed to reach the Spice Islands, but in the end, he did manage to change their spices.

In fact, Columbus and his chili peppers ultimately changed the whole spice industry. By transporting the seeds across an ocean, he showed that chili plants were like any other crop. Given the right conditions, they could flourish far outside their native range. Once this idea took hold, the trend was unstoppable. By the end of the eighteenth century, nutmeg had moved to Grenada, cloves and cinnamon had shown up in Zanzibar, and people had started planting black pepper wherever a tropical vine could climb a tree stump. Cheap product flooded the market, prices plummeted, and spices lost their exotic cachet. Though it remained a valuable en-terprise, the spice trade never again sparked wars, founded empires, or inspired voyages of discovery. But for centuries, the lust for spices shaped history, and seeds lay at the heart of it. Seeds still dominate the contents of a typical grocery-store spice aisle, but though people pinch, grind, dash, and otherwise consume them every day, few con-sider the biology behind that simple act. Why are spices spicy? As it happens, no story answers that question more completely than the story of chilies, the peppers of Columbus.

"It all comes down to seed production," Noelle Machnicki told me, and she ought to know. As the author of a doctoral dissertation titled "How the Chili Got Its Spice," Noelle has spent more time thinking about hot peppers than just about anyone. When I caught up with her, she'd recently defended her thesis and was busy holding down two different jobs at different universities in different cities. "I'm sort of living a double life right now," she admitted wearily, sip-ping from a large coffee. Noelle has dark hair, dark eyebrows, and an expressive face that can shift from wariness to warmth in an instant. When the conversation turned to chilies, all signs of fatigue disap-peared and she suddenly spoke with the enthusiasm of someone who can't wait to tell you a secret. Her work caps fifteen years of research by the "Chili Team" at the University of Washington's Tewksbury

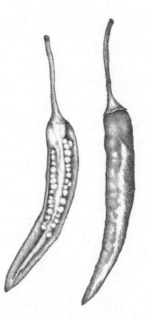

FIGURE 9.2. Chili pepper (*Capsicum* sp.). The thousands of varieties of domestic chili peppers descend from four species native to South America. In the wild, their pungency repels seed-killing fungi as well as rodents and other mammals that can't take the heat. ILLUSTRATION © 2014 BY SUZANNE OLIVE.

Lab. Taken together, their research papers epitomize how science is supposed to function: questions leading to insights, leading to new questions, until a fascinating drama lies revealed. For Noelle, it all started with a love of mushrooms.

"I'm basically a mycologist," she said, and explained how the prolific toadstools of the rainy Pacific Northwest had helped draw her from her home near Chicago. She studied them on the forested campus of Washington's Evergreen State College, and then entered graduate school to pursue a particular passion. "I'm fascinated by how fungi interact with plants," she told me—how they exchange nutrients with roots in the soil and show up everywhere from bark to flowers to the insides of leaves. So when biology professor Joshua Tewksbury asked her to help identify a fungus growing on wild chili-pepper seeds, she was all ears. At the time, Tewksbury had already followed his research on chilies from the American Southwest to the Chaco region of Bolivia, where he'd discovered a species that varied in pungency from completely mild in dry habitats to what Noelle described as "definitely hotter than Tabasco" in wet ones. Intermediate places had the two forms growing side by side, and the only way to tell the difference was to taste them—sometimes hundreds in a day. Luckily for Tewksbury, he'd found the ideal collaborator: a mycologist who liked spicy foods. "I do tolerate chilies better than the average person," she allowed. But when I pressed the question, she laughed, and confessed to keeping a bottle of hot sauce in her desk drawer at work. "Josh does too!" she added.

The Bolivian chilies presented a rare opportunity. They seemed to preserve that key moment in time when pungency was just evolving. "We know the first chilies weren't hot," Noelle said firmly, and explained that all modern species, no matter how spicy, descended from a mild common ancestor. Whatever ecological dilemma caused that distinctive hotness to evolve appeared to be ongoing in Bolivia, where some chilies had made the switch and some hadn't. If Noelle and the rest of the team could figure out what was going on, they would indeed know how, and why, the chili got its spice. Chemically, the answer was already in the bag.

Scientists long ago traced the pungency in chilies to the presence of *capsaicin*, a compound produced in the white, spongy tissue that surrounds the seeds. It's what experts call an *alkaloid*, a type of chemical that may be more familiar than you think. Alkaloids all share a similar nitrogen-based structure, a set of building blocks that plants have arranged and rearranged into more than 20,000 distinct combinations. The nitrogen matters because it's a vital nutrient that plants also need for growth, so they don't use it on alkaloids without a purpose. Usually, that purpose amounts to some form of chemical defense. And because plants usually need to defend themselves against animals, alkaloids almost always have an effect on people, too. They can be spicy, like capsaicin, but that's just the beginning. Even a short list of common alkaloids includes some of the world's most recognizable stimulants, narcotics, and medicinals, from caffeine and nicotine to morphine, quinine, and cocaine. In Bolivia, however, few mammals seemed interested in chili peppers, even the mild ones. To Noelle, that made the fungus growing on the seeds look all the more suspicious.

"A fungal seed pathogen is the strongest kind of selection pressure," she explained. "Seeds are progeny—a direct link to fitness." In other words, if fungi were killing the seeds of mild chilies, it would give the plants a powerful reason to develop some kind of chemical response. After all, there's hardly a stronger evolutionary imperative than the life or death of offspring. In an elegant series of experiments, Noelle showed that fungi did indeed kill a large portion of the seeds they infested, and that pungent seeds were significantly more resistant than mild ones. Capsaicin slowed or stalled the growth of a wide range of fungi, both in the wild and in Petri dishes back at the lab, strongly suggesting that it evolved for just that purpose. But her success only raised another question. Why weren't all the chilies hot? If capsaicin was such a great idea, then why did some plants keep producing chilies as mild as an apple?

Solving that puzzle takes us back to the great square dance of coevolution, the same process of give-and-take that led to strong rat teeth and thick nutshells. In this case, the struggle was invisible, but no less imperative. Noelle's research showed that both chilies and fungi respond to one another—plants produce more capsaicin as the fungi become resistant, and vice versa. "I think of it as a coevolutionary arms race," she summarized, but running that race had steep costs for both sides. For a fungus to withstand capsaicin, it gave up the ability to grow quickly—a distinct disadvantage anywhere except inside a pungent chili pepper. For the plants, making capsaicin interfered with their ability to retain water, leading to lower seed production in dry weather. What's more, it took energy away from the woody material in seed coats, making the seeds more vulnerable to predation by ants. These are serious drawbacks that only made sense under certain conditions, a reminder that the results of coevolution involve more than which partners are dancing. It also depends on where the dance takes place.

Bolivia's Gran Chacò region stretches from arid savannahs and cactus patches to wet forested hillsides near the borders of Paraguay and Brazil. By sampling chilies across 185 miles (300 kilometers) of mixed terrain, Noelle and her team quickly found a pattern. "In areas of high rainfall, all the chilies are hot," she told me. "But as rainfall decreases, so does pungency." For chilies growing in wet forests, where fungi and the insects that move them from fruit to fruit are common, investing in spiciness was a clear advantage. But in arid environments, fungi don't grow nearly as well, and the potential for water stress and low seed production made pungency a burden. This dynamic of pros and cons put the evolution of pungency in context—a balance between rainfall, insects, fungi, and the physical costs of producing capsaicin. It also helps explain how a change in climate, range, or habitat might have led the ancestors of domesticated chilies to lose their mild forms entirely. When life gets wet and moldy, chilies retaliate with heat.

Most spices will never receive the level of scrutiny that Noelle and her colleagues have given to chilies, but the capsaicin story illustrates a general pattern about the pathway to spiciness. Similar research may one day unravel the mystery behind the myristicin in nutmeg and mace, or the piperine that puts the punch in black pepper. What we perceive as spiciness develops in an intricate coevolutionary dance between plants and their adversaries. Without those relationships, world cuisine would be almost universally bland. This raises a question worth thinking about: Why is it that we add seeds, bark, roots, and other plant parts to flavor meat dishes, and not the other way around?

From pepperoni and pepper steak to pork vindaloo, the zest in our favorite meat recipes always comes from the spices, not from the meat. There is a fundamental biological reason for this. Meat isn't spicy because meat can move. When a chicken, a cow, a pig, or virtually any other animal is attacked, its capacity for motion gives it a wide range of options: run away, take flight, climb a tree, slither into a hole, or stand and fight. Plants, on the other hand, are stationary. Their lot in life is to stay put and endure, a situation tailor-made for the evolution of chemicals. If you can't flee or fight back physically (beyond the occasional spine or thorn), it makes perfect sense to repel attackers with alkaloids, tannins, terpenes, phenols, or any of the many other compounds invented by plants. It's true that insects also boast a wide array of chemical defenses, but they often get them from the plants they eat. Some frogs and newts manufacture poisons, too, and there is at least one species of toxic bird. But the only notable exception to the bland-animal rule comes from the ocean floor, where bryzoans, sponges, anemones, and a range of other creatures spend most of their lives glued to rocks, as stationary as plants. Thousands of marine alkaloids have been isolated from these animals, though it remains to be seen if any of them will prove tasty sprinkled on fajitas, souvlaki, or chicken tikka.

Before the end of our conversation, I asked Noelle what remained to be learned about capsaicin and chili peppers—what were she and her colleagues working on now? The discussion immediately veered into whole new topics, each one as potentially groundbreaking as Noelle's dissertation. The birds that disperse chilies, for example, appear to be utterly immune to their pungency. They gobble the fruits at will, and the seeds pass through unharmed, or even enhanced, since it appears that moving through a bird helps clean them of fungi. Capsaicin also slows the digestion of birds, forcing them to carry the seeds farther. Noelle told me that the insects moving fungi from fruit to fruit may be chili pepper specialists, and she talked about a student studying how ants differentiate pungent seeds from mild ones. Then she mentioned that someone had recently discovered a fungus capable of making its own capsaicin—though why on earth it would want to is still anyone's guess. But perhaps the most fascinating line of study has to do with the effect of capsaicin on mammals, which is, after all, the reason that Christopher Columbus packed his hold with chilies, and why they quickly found welcome in spice drawers around the world.

When capsaicin touches the human tongue, sinuses, or other sensitive areas, it produces what chemists describe as "a sensation of intolerable burning and inflammation." Chefs and fans of hot sauce might describe it differently, but the cause is the same: a chemical sleight of hand that confuses the body's natural system for detecting heat. Normally, burn sensors in the skin activate only above 109°F (43°C), a temperature that can begin causing physical damage to cells. When you scald your mouth on hot soup, for example, the pain you feel is an honest use of this system. Biting into a pungent chili, however, triggers that response at any temperature. Capsaicin molecules target those same burn receptors and open the floodgates, tricking the body into the kind of pain and rush of endorphins that would normally accompany a serious wound. From the brain's perspective, the mouth is on fire. The feeling may last for seconds, minutes, or even longer at high doses, but eventually the capsaicin dissipates and the body recognizes that no harm has been done.

For people, this sensation can be enjoyable, the culinary equivalent of a roller-coaster ride or a horror movie—scary, without actually being dangerous. According to some studies, exhilaration from the endorphins peaks only after the burning sensation fades, which raises the paradoxical possibility that we eat chili peppers precisely because it feels so good to stop eating them. Noelle likes spicy food well enough to keep hot sauce on hand at all times, even at the office. But she thinks people developed a taste for pungency only out of necessity, and that chilies entered the human diet for another purpose. "Adding small quantities to food is a pretty good preservative," she said, noting that capsaicin deters a whole range of microbes in addition to fungi. It's telling that chilies—and so many other spices—were domesticated in the humid tropics, where meat and fresh vegetables spoil quickly. For thousands of years before refrigeration, having a burning tongue was small price to pay for thwarting mold and harmful bacteria. If Noelle is right, then people started eating capsaicin for the very same reason it evolved: to ward off the fester of fungus and rot.

Without the need to preserve meat stews or pots of beans, no other mammals have developed the chili pepper habit. They feel the same burn that we do, but to them, pain is simply pain. So while pungency may have gotten its start fighting fungi, it's also extremely good at deterring rats, mice, voles, peccaries, agoutis, and all the other mammals that would otherwise happily devour a chili seed. Where these gnawers are common, that's an important evolutionary advantage for the chilies, and it almost certainly played a role in determining why pungency became dominant in so many chili species. It also creates a brilliant dispersal strategy: repel the animals that chew and destroy your seeds, leaving more available for the birds, whose pain receptors don't respond to capsaicin, making them physically incapable of feeling the heat.

When I said goodbye to Noelle, my head was still swimming with chili questions. But that's the way with science—new information simply feeds the curiosity. The complexity of the chili pepper's story explains not only how seeds can become spicy, but why spices have so many uses in addition to seasoning. If they evolved to interact with everything from bacteria and mushrooms to squirrels, it's no wonder that people find spices useful in a lot of situations. In Columbus's day, they certainly found their way into food, but they also served as popular medicinals, aphrodisiacs, preservatives, and oblations. (Contrary to popular myth, exotic spices were never used to cover up the taste of rotting meat. They cost a fortune and signified status—the people who bought them could easily afford fresh, high-quality ingredients.) In modern times, things haven't really changed all that much. Capsaicin from chilies—to take just one example—forms the basis of everything from arthritis creams and weight-loss pills to condom lubricants, bottom paint for boats, and the self-defense spray marketed as Mace. Olympic show-jumpers have been disqualified for rubbing it on the legs of their horses, and wildlife rangers in Africa fire it from drones to herd elephants away from poachers. But in China, people use capsaicin for something that most of us associate with a different seed product, one perhaps even more famous than chili peppers.

Chairman Mao Zedong promoted an austere lifestyle with simple peasant foods, but he harbored a famous love of chilies. Even while living in a cave, he ordered them baked into his bread, and reportedly ate whole handfuls to boost his energy while working late at night. Now, police officers in Mao's native Hunan region regularly distribute hot chilies to sleepy drivers in an effort to reduce traffic accidents. For most night owls, however, the stimulant of choice comes in liquid form, extracted from the seed of an African shrub. Like spices in their heyday, it has spawned vast fortunes, influenced world events, and inspired at least one sea voyage worthy of an adventure story.

CHAPTER TEN

The Cheeriest Beans

If I can't drink my bowl of coffee three times daily, then in my
torment I will shrivel up like a piece of roasted goat!

> —Johann Sebastian Bach and Christian Friedrich Henrici,
> *Schweigt Stille, Plaudert Nicht*, aka *The Coffee Cantata* (c. 1734)

In the year 1723, a French merchant ship sat becalmed halfway across the Atlantic Ocean. For over a month, she drifted with the currents, sails loose and flapping, waiting for a steady breeze. More than two hundred years had passed since Columbus made the same journey, and transatlantic travel was now a matter of course. But sometimes the fate and consequence of a voyage still hinged on seeds. By some accounts, the drifting ship had already faced a troubled passage—outrunning a deadly storm off Gibraltar, and narrowly avoiding capture by Tunisian pirates. Now, stuck in that windless equatorial zone known as the doldrums, the ship had run so low on fresh water that the captain ordered strict rationing for crew and passengers alike. Among those travelers, one gentleman felt particularly parched, because he was sharing his small allotment with a thirsty tropical shrub.

"It serves no purpose to go into the details of the infinite care I had to provide that delicate plant," he wrote, long after the wind picked up and the ship docked safely at the Caribbean island of Martinique. And long after the descendants of his spindly sapling were well on their way to changing economies throughout Central and South America. The plant, of course, was coffee, but just how a young naval officer named Gabriel-Mathieu de Clieu got his hands on it remains a matter of debate.

In one version of the story, de Clieu and a band of masked colleagues scaled the walls of the Paris botanical gardens, broke into a greenhouse, uprooted a young coffee tree, and fled into the night. Most historians regard this reported chain of events with suspicion, but no one disputes its location. In the early eighteenth century, the only coffee plant in all of France resided at the Jardin Royal des Plantes. It was a large, healthy specimen that had been given as a token of esteem to King Louis XIV from the city of Amsterdam. De Clieu described his coffee plant as small, "no larger than the slip of a pink," so it must have been either a cutting or a seedling grown from the Sun King's tree. The royal gardeners had been trying to propagate coffee as a horticultural rarity, but they may not have recognized its huge economic potential. Having traveled widely, de Clieu knew that people in the West no longer regarded coffee as an exotic novelty, a beverage of Turks and Arabs. It was becoming a daily staple from London to Vienna to the colonies, consumed not only in cafes and coffeehouses, but in people's homes. Dutch plantations on Java dominated the world market so completely that the word *java* would soon become synonymous with the drink itself. Bringing coffee to Martinique, where de Clieu owned a large estate, promised to break the Dutch monopoly, bolster the French Empire, and earn de Clieu a tidy profit in the process.

"Immediately upon arrival in Martinique," he later recalled in a letter, "I planted . . . the precious shrub, which had only become dearer to me through the dangers it had known." Those dangers included more than a water shortage. De Clieu's correspondence revealed other details: a jealous fellow passenger who repeatedly tried to steal the sapling, and succeeded in tearing off a branch; the

FIGURE 10.1. French naval officer Gabriel-Mathieu de Clieu famously shared his water ration with a small coffee tree while becalmed crossing the Atlantic in 1723. Cuttings and seeds from that lone tree helped him found coffee plantations throughout the Caribbean and perhaps as far as Central America and Brazil. Anonymous (nineteenth-century). WIKIMEDIA COMMONS.

constant guard and fence of spikes needed to protect the little plant once it reached his estate; and the hinted possibility that he had obtained his tree not by theft but through romance, that is, by charming a "Lady of High Standing" at the French court. Centuries later, truth and embellishment are impossible to unravel, but in any form de Clieu's exploits show how far people are willing to go for a good cup of coffee. When his precious shrub finally did bear fruit, all that persistence paid off handsomely. De Clieu shared seeds and cuttings with neighboring plantations, and within a few decades Martinique boasted nearly 20 million highly productive trees.

Though little remembered today (his Wikipedia entry runs fewer than 250 words), Gabriel-Mathieu de Clieu once enjoyed a certain celebrity among coffee drinkers. English poet Charles Lamb paid tribute to him in 1810 with a verse that began:

Whenever I fragrant coffee drink,
I on the generous Frenchman think,
Whose noble perseverance bore,
The tree to Martinico's shore.

De Clieu wasn't the only person to carry coffee across the Atlantic, but people like Lamb gave him credit for every coffee tree from Martinique to Mexico to Brazil, a region that now accounts for over half of world production. That claim exaggerates de Clieu's role, but the Frenchman did get one thing exactly right: demand for coffee was rising. Since de Clieu's day, global coffee consumption has skyrocketed. As the 1940 Inkspots classic "Java Jive" pointed out, people are fond of buying "a cheery coffee bean—boy!" That fondness has transformed the seeds of a shrubby African tree into the world's second most traded commodity. Only oil futures generate more annual revenue. For the estimated 1 billion to 2 billion daily partakers, myself included, the ritual of buying, brewing, and drinking coffee rarely includes a basic question: Why do we bother? If it comes up at all, this query elicits a quick response: caffeine, the mildly addictive stimulant found in abundance in coffee beans. But that answer only invites another question: Why is coffee caffeinated in the first place?

If Charles Lamb had truly wanted to say thanks for his morning cup, he should have penned an ode to various insects, slugs, snails, and fungi. Instead of couplets like, "The islanders his praise resound / Coffee plantations rise around," he might have sought a rhyme for the way caffeine slows the heart rates of snails, or how slugs respond with what one research group called "uncoordinated writhing." The poem should have mentioned hornworms and shot-borer beetles, whose larvae wither at the merest hint of caffeine, and it might have also explained how caffeine slows the growth of fungal pests, from common root rot to witch's broom. But poets don't think about larvae and fungi when they make a pot of coffee—nobody does. The fact remains, however, that we wouldn't be drinking the stuff without them.

"Caffeine Is a Natural Insecticide," trumpeted a headline in the *New York Times*, soon after researchers published an early account of its effects. The story was brief, but singled out mosquitoes as particularly susceptible. In fact, caffeine is so effective, and against such a broad range of pests, that coffee wasn't the only plant to think of it. The seeds of at least three other tropical trees also contain caffeine: cacao, guaraná, and kola nut. Like coffee beans, these can all be ground and mixed with water to make beverages—hot cocoa, the guaraná sodas of Brazil, and a host of drinks marketed as kolas or colas, including the original versions of Coke and Pepsi. Caffeine also occurs in the leaves of tea and in a type of South American holly known as maté, pretty much rounding out the list of humanity's favorite stimulating liquids. It seems that wherever caffeine turns up in nature, people aren't far behind, holding out our mugs, gourds, and samovars.

Like capsaicin, caffeine is an alkaloid. Producing it requires precious nitrogen that might otherwise be used for growth, so coffee trees make the most of their investment through what amounts to a caffeine-recycling program. They manufacture it only in the most vulnerable tissues, and later transfer that caffeine to the most important place of all, the seeds. The process starts inside young leaves, where caffeine helps fend off insects and snails that prey on tender foliage. But as those leaves grow and toughen, the plant withdraws much of that caffeine and redirects it to protect flowers, fruits, and the developing seeds. The fruit, a reddish berry, also produces caffeine, much of which diffuses inward to the pair of seeds nestled inside. And those seeds not only receive caffeine, they make more, resulting in a concentration potent enough to fend off all but the hardiest attackers. In total, more than nine hundred species of insects and other pests target coffee trees, so it's logical to assume that caffeine evolved in response. But just as historians can't agree on the particulars of Gabriel-Mathieu de Clieu's story, scientists can't all agree on the evolution of caffeine. It may be a good pesticide, but that's not the only thing it's used for.

Coffee plants manufacture caffeine in various places, but once it reaches the seeds it stays put, bound up in the cells of the endosperm. That's good news for coffee drinkers, but a mixed blessing for the seeds, because caffeine does more than fend off attackers—it also prevents germination. The same chemistry that kills beetle larvae and makes slugs writhe interferes with cell division in plants. We touched on this dilemma earlier, but it bears repeating: to sprout successfully, coffee needs to get its tiny root and shoot away from the caffeinated part of the bean. It accomplishes this by imbibing water rapidly, flooding premade cells that swell and push the growing tips outward. Only after they make their escape from the bean can cell division and true growth begin. But once that takes place, something even more interesting occurs. As the seedling gets bigger, caffeine leaks from the dwindling endosperm and spreads into the surrounding soil, where it appears to curb the growth of nearby roots and stop other seeds from germinating. In other words, coffee beans know how to kill off the competition—they release their own herbicide, clearing a tiny patch of ground to call their own. In a seed's vital struggle to sprout and get established, that's an evolutionary advantage every bit as important as warding off pests.

It's easy to understand why coffee plants would want to protect their seeds and leaves, or to give their seedlings a head start. The final theory on caffeine evolution is more surprising, but early in the morning it's one that a lot of people can relate to. It has to do with addiction. As recycled caffeine moves around inside a coffee tree, it shows up in one place that puzzled scientists for a long time: flower nectar. What is the point in putting insecticide into something designed to attract insects? Recent research on honeybees has revealed the answer. At the right dosage, caffeine doesn't repel pollinators, it keeps them coming back.

"I think it amplifies responses of neurons in their reward pathway," Geraldine Wright told me. As a professor of neuroscience at Newcastle University, Wright has made a career out of studying how honeybees think. She knows them well enough to occasionally don a "bee bikini" at public events, a live swarm of workers that covers her torso from chest to neckline. Bee brains may be simple, but they're capable of great feats of cooperation. When Wright and her colleagues trained a hive to visit experimental flowers, the bees were three times as likely to remember and return to the ones dosed with caffeine. In this case, at least, honeybee brains work just like ours do—their "reward pathways" light up when they drink caffeine. For coffee trees, producing caffeinated flowers attracts a dedicated cadre of pollinators, lined up like morning commuters at their favorite espresso stand.

When I asked Wright whether caffeine might have evolved for this purpose—whether its potency as a pesticide and herbicide was just icing on the cake—she seemed to think that was a stretch. "I'm not sure the selection pressure would be strong enough," she wrote in an email, and I could almost picture her skeptical frown. But the fact that caffeine also occurs in the flower nectar of citrus trees, and not in their seeds or leaves, suggests that it may be possible. Oranges, lemons, and limes defend themselves with volatile oils and other compounds, apparently reserving caffeine for the express purpose of manipulating bee brains.

For a discussion of seeds, divining exactly how caffeine evolved is less important than understanding what it does—it's equally effective at warding off insects and thwarting nearby plants. But the bee story is relevant, too, because no trait has had a greater impact on the history of coffee, and the cultures that drink it, than the effect those caffeinated seeds have on the human brain.

"The emotions are raised in pitch, the fancies are lively and vivid, benevolence is excited, . . . both memory and judgment are rendered more keen, and unusual vivacity of verbal expression rules for a short time." So observed a British medical journal in 1910. Modern scholars may be more restrained in their language, but their data point to the same conclusions. Drinking an average cup of coffee releases enough caffeine into the bloodstream to measurably impact the central nervous system. Neurons in the brain fire more rapidly, muscles twitch, blood pressure increases, and drowsiness fades. But just as capsaicin burns without burning, so does caffeine stimulate without actually stimulating. The boost we feel from coffee comes less from what caffeine does to the brain than from what it prevents. Experts call caffeine an "antagonist" because it interferes with the natural function of certain brain chemicals, particularly one called *adenosine*. Researchers still don't know all the things adenosine does in the brain, but there's a way to explain its basic role that will be familiar to millions of radio listeners.

For decades, Garrison Keillor's program *A Prairie Home Companion* has featured advertisements from "The Ketchup Advisory Board," a fictional industry group promoting tomato ketchup for its "natural mellowing agents." The skits feature bland characters whose behavior becomes increasingly erratic and impulsive without regular helpings of ketchup. They suddenly decide to run marathons, pierce their noses, write memoirs, or rob liquor stores. Adenosine is not ketchup—it's one of the basic biochemicals that make the body function. But in terms of brain activity, the role of adenosine can't be described any better. It's a natural mellowing agent, slowing the neurons and triggering a whole chain of events that lead eventually to sleep. Coffee drinkers feel alert because caffeine gets in the way of that process, and even reverses it—replacing adenosine and tricking the brain into speeding up when it would otherwise be slowing down. Caffeine doesn't actually give people energy; it just renders them less capable of feeling tired.

In Ketchup Advisory Board stories, mellowness always returns when the characters get their ketchup, just as brain chemistry and sleep always overwhelm the effects of caffeine in the end. But people seem to enjoy the sensation of tricking their brains into temporary liveliness, and, like bees, they seek it out again and again. And just as a few bees can lead their whole hive to caffeinated flowers, so, too, has the coffee habit altered the course of entire human societies. In the West, historians believe it helped pave the way for both the Age of Enlightenment and the Industrial Revolution that followed. And it all started with a change in the morning meal.

Advertisers know the phrase "Breakfast of Champions" as an iconic slogan used by the Wheaties cereal brand for over eighty years. To students in fraternities and college dormitories, however, it's not a "breakfast of champions" until the Wheaties are doused with a generous portion of beer. After a night of excess, this combination is promoted as a sort of hair-of-the-dog hangover cure, but the soggy result is something that few people try more than once. Bleary-eyed undergrads might be surprised to learn that people throughout central and northern Europe greeted *every day* with a version of this breakfast for more than nine centuries. Before the advent of coffee, "beer soup" was the morning staple. The standard recipe featured steaming-hot ale poured over bread or mush, with eggs, butter, cheese, or sugar added on special occasions. This concoction provided people of all ages with carbohydrates, calories, and, although the beer was usually weak, a modest buzz. In fact, the morning soup simply marked the first installment of a long, beery day. Alongside bread, home-brewed ales and other beers were part of every meal, making up a significant nutritional component of the medieval diet. As late as the seventeenth century, when coffee began to take hold, per capita beer consumption in northern Europe ranged from 156 to as much as 700 liters annually, with 300 to 400 liters considered average. Modern figures pale by comparison—Americans drink a paltry 78 liters per year, the British put away 74, and even the beer-loving Germans knock back only 107.

Into this environment of habitual tipsiness, coffee arrived as what social historians have dubbed "the Great Soberer." Instead of the mental fog brought on by beer (or wine, which was the staple in southern Europe), drinking coffee made people alert, energetic, and arguably more productive. The college-student analogy works here, too: anyone hoping to graduate figures out pretty quickly that drinking beer before classes produces a much different outcome than drinking coffee. Both are seed products, but replacing the fermented one with the stimulant has profound effects on more than just a grade point average. In Europe, the coffee transition occurred on the

heels of the Reformation, and its promise of sobriety and productivity fit neatly into the era's emerging philosophy. As one scholar put
it, coffee "achieved chemically and pharmacologically what rational-
ism and the Protestant ethic sought to fulfill spiritually and ideologically." In practice, coffee prepared both body and mind for the kinds
of indoor work becoming common in towns and cities—the jobs
of governance, commerce, and manufacturing. It's no coincidence
that the modern definitions and spellings for "coffee," "factory," and
"working class" all entered the English language in the eighteenth
century. The beverage became especially popular with workers in urban areas, and London once boasted as many as 3,000 coffeehouses,
one for every 200 inhabitants.

Like any craze, the coffee phenomenon included no small amount
of hype and hyperbole. Though it was legitimately prescribed as a
stimulant, doctors and hucksters also recommended the beverage
for any number of other ailments, from gout and tuberculosis to ve-
nereal disease. Some claims were contradictory (headache cause vs.
headache cure), and most proved false (aphrodisiac, increased intelligence), but others remain subjects of medical inquiry (antidepres-
sant, prevention of tooth decay, appetite suppressant, hypertension
remedy). The continued research interest in coffee should come as
no surprise. Coffee beans contain at least 800 other compounds in
addition to caffeine—making that daily cup, by some accounts, the
most chemically complex food in the human diet. Most of coffee's
components have never been studied, so their health effects remain
mysterious. Researchers generally agree that coffee drinkers enjoy a
reduced risk of type II diabetes, liver cancer, and, for men at least,
Parkinson's disease. But nobody has any clear idea why.

Drinking too much coffee can lead to restless nights and the kind
of nervous jitters that Johann Sebastian Bach satirized in the title of
his so-called "Coffee Cantata:" *Schweigt Stille, Plaudert Nicht*—"Be
Still, Stop Chattering." Bach himself was a famous coffee lover and
hosted regular performances of his work at Café Zimmermann, the
finest coffeehouse in Leipzig. Such gatherings exemplify the role that

coffee had begun to play socially and culturally in the eighteenth century. Because it stimulates thought and conversation in a much different way than alcohol, coffee brought people together not for revelry, but for serious conversations, meetings, and cultural events. A trip to a coffeehouse was (and still is) quite different from visiting a tavern. People not only met friends there, but also gathered to study, hear the news, play chess, and even conduct business. The shipping underwriters that frequented Edward Lloyd's establishment in London went on to form the largest insurance marketplace in the world, which still goes by the name of its coffee-vending founder. Nor is Lloyd's of London the only famous example. The Bank of New York was organized at Merchant's Coffee House; the London Stock Exchange got its start in a shop called Jonathan's; and the public sales held at coffeehouses—for everything from artwork and books to carriages, ships, real estate, and "things seized as the goods of pirates"—led to the founding of the world's two great auction houses, Christie's and Sotheby's.

To philosophers, writers, and other intellectuals, coffeehouses quickly became indispensable hubs for articulating and sharing ideas. People called them "penny universities," claiming you could get a good education by simply listening in on all the highbrow conversations. Voltaire reportedly drank fifty cups of coffee every day; he spent so much time in Paris's Café de Procope that his writing desk remains there, enshrined in a corner. Rousseau frequented the Procope as well, where he reportedly practiced his chess game against the great encyclopedist Denis Diderot. The luminaries of Samuel Johnson's Literary Club met for coffee at the Turk's Head in Soho for nearly twenty years, and Jonathan Swift was so dedicated to the St. James Coffeehouse that he had his mail delivered there. Scientists liked coffee, too, and although stories of Sir Isaac Newton dissecting a dolphin at The Grecian Coffeehouse are false, he did spend a lot of his evenings there. It was a popular destination after meetings at the nearby Royal Society (which, incidentally, began life as the Oxford Coffee Club).

Political thinkers flocked to the coffeehouses, too. Robespierre and other key figures of the French Revolution often met at Café Procope, and a young Napoleon Bonaparte once had to leave his hat there as collateral when he couldn't pay a bill. Benjamin Franklin dropped by whenever he was in town, and in London, his coffeehouse friends, "The Club of Honest Whigs," included the radical liberal Richard Price. Price's ideas had a strong effect on Franklin and other leaders of the American Revolution, proving that Charles II had been right, decades earlier, in railing against coffeehouses as centers of sedition. To imply that drinking coffee *caused* revolutions would be going too far, but it's no exaggeration to say that it caused revolutionary thinking. As a drug and a social rallying point, coffee played a role in transforming the ideals of the Enlightenment into political reality.

In putting coffee at the center of cultural and political events, Europeans had adopted more than an Arab beverage: they'd taken on an Arab way of life. For centuries before coffeehouses became fashionable in Paris and London, they had served as community gathering places throughout the Near East and North Africa. (Legend traces the origin of coffee to an Ethiopian goatherd who noticed his flock dancing on their hind legs after feeding on the beans.) As a social, nonalcoholic indulgence, coffee was well suited both to the tenets of Islam and to what scholars consider one of the world's most deeply conversational societies. The influence of coffeehouses waned in the West during the nineteenth century, but places like Cairo's Al-Fishawy cafe haven't closed their doors in over 260 years. One needn't look farther than the title of a recent academic paper to see the continuing importance of coffee in the Arab world: "Clicks, Cabs, and Coffee Houses: Social Media and Oppositional Movements in Egypt, 2004–2011." Throughout the "Twitter Revolutions" of the Arab Spring, coffeehouses served as essential physical meeting points—planning centers, places of refuge, and even makeshift hospitals. In Egypt and across the region, those are roles they've played during virtually every popular uprising for the past five centuries.

If Gabriel-Mathieu de Clieu crossed the Atlantic today, he would find the production and processing of coffee well established in the Caribbean and throughout Central and South America. But if he wanted to know about *drinking* coffee, people would probably send him to a place not far from my home, a city that has been called the coffee "Mecca" of North America. When Howard Schultz installed the first espresso machine at Seattle's Starbucks Coffee in 1983, he helped spark what can only be called a coffeehouse renaissance. The coffee bean hadn't seen such a boom in North America and Europe since the eighteenth century, and Starbucks alone now boasts over 20,000 stores in sixty-two countries. This rise hasn't occurred in a cultural vacuum. It's not surprising that Starbucks got started in the same urban center that gave the world Microsoft, Amazon, Expedia, RealNetworks, and a host of other technology companies. Coffee may have been a good match for the Age of Enlightenment, but it's an even better fuel for the Information Age—and the tech-driven, intensely indoor lifestyle it fosters. In the words of one expert, the caffeine delivered by coffee has become "the drug that makes the modern world possible."

The Internet, texting, social media, and other digital innovations have created longer workdays and an expectation of constant connectivity, a perfect environment for the stimulating effects of coffee. The popular technology magazine and website *Wired* takes its name from an insider term with overlapping meanings: fluent in the digital world, and buzzing from stimulants. The caffeinated habits of the old "computer geek" stereotype have gone mainstream, proliferating right alongside the screens we stare at, from our desktops to our laptops, tablets, and smartphones. Tea sales are up, too, and caffeine (often extracted from coffee beans) is now a popular additive to energy drinks, sodas, pain relievers, bottled water, breath mints, and—in a peculiar botanical twist—"energized" sunflower seeds. Where office workers used to expect lukewarm dregs from a percolator by the copy machine, employees at companies like Google, Apple, and Facebook now enjoy the benefit of full-scale coffeehouses on their corporate campuses, free of charge. But perhaps nothing underscores the relationship between coffee, technology, and the New Economy more than shops like the Surf Café in Seattle, or San Francisco's The Summit, where patrons actually rent desk space to work on their start-up ideas and meet with venture capitalists. This combination of coffee bar and cubicle is like a modern echo of the old Lloyd's, where insurance brokers started at the counter, moved on to tables and booths, and now occupy a fourteen-story, triple-tower skyscraper in downtown London. Coffee is helping shape the tech-driven economy in the same way. New ideas are created under its influence, and meetings in coffeehouses help bring those products to market.

To rediscover the seed at the heart of all this, I decided to visit a Seattle coffeehouse. (Like buying Almond Joy candy bars, drinking coffee as a business expense sounded like another career milestone.) But how does one choose where to go in a city with thousands of establishments licensed to brew and sell hot beverages? I talked to a friend in the coffee business, and then called around asking the following question: Where do people who work at coffee shops in Seattle go for a good cup of coffee?

Soon afterward I found myself crossing the threshold of Slate, recently crowned the Best Coffeehouse in America at the annual Coffee Fest trade show. Slate occupies a former barber shop on a side street in Ballard, one of Seattle's trendiest neighborhoods. (Coincidentally, it's just down the hill from where my Norwegian great aunts Olga and Regina once lived, in an era when Ballard was a Scandinavian enclave better known for pickled herring than espresso.) The ambiance inside was conspicuously spartan. A vintage record player in the corner hummed with jazz, but otherwise there were no distractions. Spare gray walls, a tidy counter, and simple bar stools put all the emphasis on the coffee. In the wrong hands, this setup might have seemed contrived, but the people at Slate overwhelmed any hint of pretension with friendliness, and with an enthusiasm for coffee as unadorned as the walls.

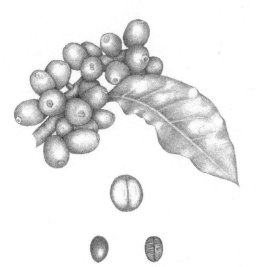

FIGURE 10.2. Coffee (*Coffea* spp.). Beloved for their stimulating caffeine and complex flavor, the seeds of these small African trees have become the world's most traded commodity. The berry-like fruits pictured at the top and in cross section each contain two seeds that swell and darken when roasted (below). ILLUSTRATION © 2014 BY SUZANNE OLIVE.

"I'll put you right over here," said co-owner Chelsey Walker-Watson, greeting me at the door with a smile and a handshake. She sat me at the counter between two people who were also holding notepads, and for a terrible moment I thought that they, too, must be writing books about seeds. But then Chelsey introduced them as new employees and explained that I would be sitting in on a training session. So for the next three hours I remained there at the counter—brewing coffee, drinking coffee, talking coffee, and learning what it takes to be a barista at the hippest coffeehouse in the country.

"Basically, I got a job at Peet's because my boyfriend wanted free coffee," Chelsey admitted, when I asked how she got started in the business. Petite, with hair the same dark shade as the frames of her glasses, she had a self-deprecating style that belied the obvious success of her career. I didn't ask whether the boyfriend had lasted, but the coffee certainly had. She'd spent a decade rising through the ranks at Peet's—the third-largest specialty chain in the country—before leaving to start Slate. And now, within a year of opening, the shop was already winning national awards. The Slate approach gets back to basics, focusing on coffee beans as the seeds of a plant, and recognizing that differences in growing conditions—soil, elevation, rainfall—can have a distinct impact on how those beans come out. They can vary not only in size, color, and density, but also in their chemistry, since the pests they face in a place like Vietnam will be very different from those found in Ethiopia, Colombia, or Martinique. Where most coffeehouses strive for uniformity from cup to cup, the Slate team tinkers with every roast and brew to highlight any potential difference in flavor.

"It's like making toast," explained head barista Brandon Paul Weaver. "White bread and whole wheat bread are very different, but if you burn them they taste exactly same." The trick lies in roasting the beans just enough to make toast, but not so much that they lose their uniqueness. "They can't be too raw, either," he qualified, making a face. "Raw beans just taste like grass."

After all the buildup, I wasn't sure what to expect when Brandon handed me the first small cup of the day. But one sip confirmed that Slate's coffee wasn't like anything I brewed at home. It tasted like some kind of intense herbal infusion—coffee, but with strong hints of citrus and blueberry. "What do you think?" Brandon asked eagerly, drinking from his own small cup. "Are you getting the jasmine notes?"

Tall and lanky, Brandon wore his dark curls long, with a gravity-defying straw hat perched haphazardly on the back of his head. He took over the training session as Chelsey got busy with customers, speaking in rapid bursts about grind texture, water temperature, and saturation point. Brandon brewed each cup separately, using hot plates, scales, and large beakers like something out of a chemistry lab. The recipe that won him top barista honors at the Northwest Brewers Cup competition was recently posted online: "19.3 grams coffee (from the Limu region of Yrgacheffe, Ethiopia); Moderate-course grind on a Baratza Virtuoso grinder; 300 grams of 205-degree water; Kalita filters in a Clever Dripper; 3 minute, 15 second brew time."

This kind of attention to detail may seem obsessive, but Brandon, Chelsey, and everyone at Slate want coffee to take its place beside fine wine as a beverage of nuance. If they're successful, people will start appreciating coffee beans with the same focus given to wine grapes, recognizing varieties, appellations, and the rewards of a good harvest. This approach to coffee appreciation is new, and it seems to be catching on with serious connoisseurs. In contrast, Slate's goal for their coffeehouse is very traditional: they want it to be a place for conversation.

"I'm interested in what the coffee can facilitate," Brandon said, and talked about watching great interactions develop among total strangers seated at the counter. As if to demonstrate his point, our little training session soon drew a small crowd of onlookers, serious coffee enthusiasts who wanted to try Slate's methods at home. Or, for the fellow standing behind me, on the job. He was a barista at Toast, another Ballard coffeehouse. "I just got off work and thought I'd stop in for a cup on my way home," he said, with no trace of irony. The crowd ranged from skinny teenagers to a retired couple, and even a tourist from Georgia who'd read about Slate on a coffee blog. At one point, people were so intent on watching Brandon pour that no one noticed the record player was skipping, endlessly repeating a fast Benny Goodman lick on the clarinet. It was some kind of ascending jazz arpeggio, the perfect musical accompaniment to coffee—*up, up, up!*

After three hours of steady coffee tasting, my brain felt like it was starting to skip, too. I could picture a swarm of caffeine molecules up there, thumbing their noses at the adenosine. Driving away, it occurred to me that in all of our conversation, one topic hadn't come up: decaf. To aficionados like the crew at Slate, taking the caffeine out of coffee defeats the purpose and spoils the taste, but decaf still makes up 12 percent of the world market. The process usually involves solvents or complex steam and water baths, but there is a holy grail out there for decaf drinkers. Among the hundred or so species of wild coffee, a handful in East Africa and Madagascar lack caffeine naturally. Their ancestors branched from the coffee family tree sometime before caffeine evolved, and they've never developed the knack. Domesticating one of those species holds the promise of a full-flavored decaffeinated coffee straight from the bean, with no processing required. In today's market, that translates to a $4 billion idea, and plenty of plant breeders have tried it. But just because a coffee tree lacks caffeine doesn't mean it lacks pests. The decaf species face the same kinds of attackers as any other coffee, and they've developed their own suite of chemical defenses in lieu of caffeine. Unfortunately, for every variety studied to date, that chemical cocktail makes the beans disgustingly bitter. The quest for a natural decaf continues, but so far it has yet to produce a drinkable cup.

Late that night, still fidgeting and staring at the ceiling, I found myself wishing the decaf researchers Godspeed. I slept eventually, but it was a good reminder that plants don't put alkaloids like caffeine into their seeds for our pleasure. They're meant to be toxic, as indeed they are to a good many insects and fungi. It's even possible for a person to die from caffeine overdose, though one study suggests it would take 150 cups of coffee, drunk all at once, to do the job. Poisoners and assassins know that there are far more lethal options available, and, not surprisingly, many of them also come from seeds. In fact, the most notorious assassination of the Cold War revolved around three things: a bridge, an umbrella, and a bean.

CHAPTER ELEVEN

Death by Umbrella

*If you drink much from a bottle marked "poison," it is almost
certain to disagree with you, sooner or later.*

— Lewis Carroll,
Alice's Adventures in Wonderland (1865)

I n novels, thick fog always envelops the city of London just when
something dramatic is about to happen. It hides the robberies and
the kidnapping in *Oliver Twist*. It shrouds the arrival of Dracula
when he comes for Mina Harker, and Sherlock Holmes watches it
swirl down the street before the fateful events in *The Sign of Four*.
But on September 7, 1978, light morning showers had given way to
sunshine by the time Georgi Markov parked his car and began walking toward the Waterloo Bridge. Had it been foggy, Markov might
have left his windbreaker in the closet and donned an overcoat, or at
least a pair of heavier trousers. Either one could have saved his life.

At home in Bulgaria, Markov's novels and plays had made him a
famous literary star, someone who mingled with the social and political elite. He'd even gone on hunting trips with the president. Since
defecting to the West, that insider knowledge had helped him craft
accurate and scathing commentaries on repression behind the Iron
Curtain. He hosted a weekly show on Radio Free Europe and also

worked at the BBC, where he was heading on that fateful afternoon. Markov knew that speaking out put him at risk, and he'd even received the occasional death threat. But he was a relatively minor figure—no one expected him to be the target of a plot, let alone one that would become the Cold War's most infamous assassination. And no one could have predicted the murder weapon, something so preposterous even his widow found it hard to believe.

Walking past a bus stop on the south side of the bridge, Markov felt a sudden jab in his right thigh and turned to see a man bending over to pick up an umbrella. The stranger mumbled an apology, hailed a nearby cab, and disappeared. When he reached his office, Markov noticed a spot of blood and a tiny wound on his leg. He mentioned it to a colleague, but then dismissed the incident from his mind. Late that night, however, his wife found him stricken with a sudden and violent fever. He told her about the stranger at the bus stop, and they began to wonder—could he possibly have been stabbed by a poison umbrella? What had really happened was even more bizarre.

"The umbrella gun was invented by the KGB equivalent of Q's lab," Mark Stout told me, alluding to the fictional workshop for spy gadgetry made famous in James Bond movies. But while exploding toothpaste and flame-throwing bagpipes play well in Hollywood, exotic weapons are a rarity in real-life espionage. "It's almost always low-tech," Stout went on. "Somebody shoots somebody, or a bomb goes off. At the time, the umbrella gun, and the tiny pellets it fired, were a substantial feat of engineering."

I called Mark Stout about the Markov case because for three years he had held the position of chief historian at the International Spy Museum. That job title must have looked great on a business card, but it also gave him access to a working copy of the umbrella gun, built by a veteran of the same KGB lab that created the original. The replica features prominently in a section of the museum called "School for Spies," where it's displayed alongside another KGB invention, the single-shot lipstick pistol. By the time I spoke with him, Stout had moved on to a more traditional academic post, but he still showed an obvious enthusiasm for the world of secret agents. "The umbrella used compressed air, exactly like a BB gun," he explained eagerly. I could hear the squeak of his desk chair over the phone, and pictured him rolling around his office, pausing to lean back in thought. "But it was designed for extremely short range, an inch or two maximum. In Markov's case, they literally pressed the tip against his leg before firing."

For pathologists working in 1978, however, there were no spy museums or historians to turn to. Their patient soon died in a London hospital from what appeared to be acute blood poisoning, but they had no logical explanation for his symptoms. The autopsy did note an inflamed pinprick in his thigh, but it looked like an insect sting, not a stab wound. And the mysterious pellet lodged inside was so minute that technicians dismissed it as a blemish on the X-ray film. The investigation might have stopped there if another Bulgarian dissident hadn't come forward with a similar story. He'd been attacked near the Arc de Triomphe in Paris, but had recovered after a short illness. This time, doctors paid attention to his account of a painful jab, and they soon recovered a tiny platinum sphere from the small of his back. Because he'd been wearing a heavy sweater, the pellet hadn't penetrated beyond a layer of connective tissue surrounding the muscle, and most of its poison had failed to disperse. London's coroner immediately reexamined Markov's body, recovered an identical pellet from the wound in his leg, and came to a famously circumspect conclusion of foul play: "I cannot see any likelihood of this being an accident."

To the public, Markov's murder made the fantasy world of James Bond a sudden reality—in the same year, *The Spy Who Loved Me* became one of the highest-grossing British films of all time. To investigators, the case left two glaring questions unresolved: Who was the man with the umbrella, and—something British Intelligence and the CIA were keen to find out—what kind of poison could kill someone with such a tiny dose? The first question remains unanswered. Soviet defectors later confirmed that the KGB provided the umbrella and pellets to the Bulgarian government, but critical details remain hazy, and no one has ever been arrested for the crime. Resolving the poison puzzle, however, drew a unanimous opinion from an international team of pathologists and intelligence experts. They arrived at their conclusion after weeks of careful forensic analysis, with contributions from pharmacologists, organic chemists, and one 200-pound (90 kilogram) pig.

The first challenge lay in determining exactly how much poison had entered Markov's body. Measuring less than a twentieth of an inch (1.5 millimeters) in diameter, the pellet removed from his thigh contained two carefully drilled holes with a total capacity estimated at sixteen millionths of an ounce (450 micrograms). (To put that into perspective, press a ballpoint pen lightly onto a piece of paper. The tiny ink-speck it leaves behind is the size of the pellet—seeing the holes would require a microscope.) Simply knowing that dosage narrowed the possibilities to a handful of the world's deadliest compounds. The team immediately ruled out bacterial agents like botulin, diphtheria, or tetanus, all of which would have triggered telltale symptoms or immune reactions. Radioactive isotopes of plutonium and polonium didn't fit the bill, either—they can be fatal, but their victims take a much longer time to die. Arsenic, thallium, and the nerve gas sarin weren't nearly powerful enough, and while cobra venom might have produced a similar reaction, it would have required at least twice the dose. Only one group of poisons could have caused Markov's deadly combination of symptoms so quickly: the poisons found in seeds.

For thousands of years, executioners and assassins have turned to seeds in search of ways to do in their victims. The plant kingdom in general offers a vast selection of toxins, but seeds offer the advantages of easy storage and high potency. They're the most poisonous part of the hemlock plant that killed Socrates, as well as the white hellebore suspected of knocking off Alexander the Great. Strychnine trees bear seeds nasty enough to earn the nickname "vomit buttons," and their poison has figured in the murders of everyone from a Turkish president to the young women targeted by Victorian serial killer Dr. Thomas Cream. In Madagascar and Southeast Asia, hundreds of deaths every year are attributed to the nuts of a salt-marsh species known simply as the "suicide tree." The murderous potential of seeds was not lost on William Shakespeare when he needed a convincing concoction to pour into the ear of Hamlet's father. Most scholars agree that his "leperous distilment" must have been an extract of henbane seeds, just as mystery fans know that Arthur Conan Doyle modeled the "devil's foot" that nearly killed Holmes and Watson on the deadly calabar bean of West Africa. These plants all rely on alkaloids to provide their poisons, but investigators in the Markov case quickly narrowed in on a toxin more unusual, more deadly, and more difficult to trace. It's something the Castrol Motor Oil Corporation inadvertently hit on the head with their company motto: "It's More Than Just Oil."

Castrol got its start, and its name, by formulating engine oils from the seeds of the castor bean plant, a shrubby African perennial related to spurges. Castor beans store most of their energy as a thick oil that boasts a rare ability to maintain viscosity at extreme temperatures. (Although Castrol now makes a range of petroleum-based products, bean oil remains the lubricant of choice for high-performance racecars.) But the beans contain something more—a peculiar storage protein called *ricin*. Chemists know ricin for the odd, double-chain structure of its molecules. In a germinating seed, those molecules break down like any other storage protein, providing nitrogen, carbon, and sulfur to fuel rapid growth. But inside an animal—or a Bulgarian dissident—their odd structure gives them the ability to penetrate and destroy living cells. One chain pierces the surface while the other detaches inside and wreaks havoc on the ribosomes—small particles essential for translating the cell's genetic code into action. (In biochemistry, this puts ricin into a group called "ribosome inactivating proteins," known by the fitting abbreviation RIPs.) Dispersed through the bloodstream, ricin sets off a wave of

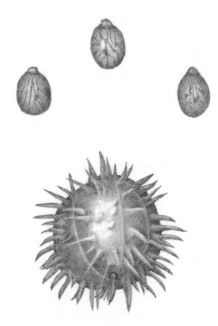

FIGURE 11.1. Castor bean (*Ricinus communis*). Beautiful enough to be sought after by jewelry makers, the mottled seeds of the castor plant contain a valuable oil as well as ricin, one of the world's deadliest poisons. The spiny, protective capsule bursts upon drying, hurling individual beans as far as thirty-five feet (eleven meters) from the mother plant. ILLUSTRATION © 2014 BY SUZANNE OLIVE.

cell death so unstoppable that even scientific journals describe it with something like awe: "one of the most lethal substances known," "one of the most fascinating poisons," or simply, "exquisitely toxic." As if to add insult to injury, castor beans also contain a potent allergen, so that, while dying, one can reasonably expect the further indignities of violent sneezing, a runny nose, and a painful rash.

Theoretically, the pellet from Markov's leg could have held enough ricin to kill every cell in his body many times over. But investigators had precious little evidence to go on. He died too quickly for any recognizable antibodies to develop, and even though ricin was known to be deadly, documented poisonings were extremely rare, and there was no clinical description of the symptoms. So the pathologists decided to stage a test. They obtained their own batch of castor beans, refined a dose of ricin, and injected it into an un-suspecting pig. Within twenty-six hours the pig died in the same horrible manner as Markov. "The . . . animal defense people would be horrified," a doctor on the case observed, but it emerged later that Bulgarian scientists had been even more brutal. They had fine-tuned the dose destined for Markov after testing a smaller amount on a prison inmate, who survived. When they worked out a quantity that would reliably kill a full-grown horse, they put the plan into action.

Georgi Markov's murder shone a bright media spotlight on the homicidal potential of seeds. Criminal elements took note, and ricin continues to surface as a bioterror weapon of choice. Anonymous letters tainted with it have been sent to the White House, the US Congress, the mayor of New York, and various other government offices in recent years, sometimes closing down mail-processing fa-cilities for weeks. When London police raided a suspected Al Qaeda cell in 2003, they confiscated twenty-two castor beans, a coffee grinder, and enough chemistry equipment to perform a simple ex-traction. (Their haul also included quantities of apple seeds and ground cherry pits, both of which contain traces of cyanide.) Seed poisons retain their appeal because they are not only potent, but also readily available. When I wanted some castor beans of my own, searching the Internet quickly revealed dozens of varieties openly and legally for sale. People still grow them for their oil and as an ornamental, and the plant has become a common roadside weed throughout the tropics. With a few clicks and a credit card, I had a batch delivered to my door—beautiful, glossy things the size of a thumbnail, their smooth coats mottled with burgundy swirls. They come in shades from umber to pink and often show up in beaded necklaces, earrings, and bracelets. In fact, bright "warning" colors make a number of toxic seeds fashionable in the bead industry, from rosary pea to coral bean, horse-eyes, and various cycads. But castor beans and other poisonous seeds remain common for another reason. It's a principle that underlies the modern pharmaceutical industry, but it was also expressed perfectly in the nineteenth century by philosopher Friedrich Nietzsche and by children's author Lewis Carroll.

People remember Nietzsche primarily for his views on religion and morality, but he also coined the maxim, "What does not kill me, makes me stronger." He meant it as a general comment on life, but this phrase also describes a truth about seed poisons. Lewis Carroll made the same point when his most famous character, Alice, cautioned against "drinking much" from a bottle marked "poison." By including the word "much," Carroll implied that drinking "a little" from such a bottle wouldn't be disagreeable at all, and might even do a person some good. Time and again, that is exactly the case with poisonous seeds. In doses short of deadly, many of those same toxins can be used medicinally—vital treatments for some of the world's most serious diseases. For Alice, the bottle in question contained not poison but a shrinking potion, preparation for the next escapade in her continuing adventures in Wonderland. Nietzsche's case seems more significant. He wrote his famed dictum shortly before suffering a mental collapse that scholars now interpret as the onset of brain cancer, one of the illnesses now being treated with seed extracts.

In the language of poisons, ricin is known as a *cytotoxin*—a cell killer. Along with similar compounds from the seeds of mistletoe, soapwort, and rosary pea, it shows great promise for assassinations at a much smaller scale: the targeted killing of cancer cells. By attaching these "RIP" proteins to the antibodies fighting a tumor, researchers have successfully attacked cancer in laboratory tests, clinical trials, and, in the case of mistletoe extracts, tens of thousands of patients. The challenge, of course, is twofold: finding the right dosage, and making sure the poisons don't diffuse to other parts of the body.

Whether or not ricin will become a widespread cancer treatment remains to be seen. If it does, it will join a long list of other seed and plant-based curatives dating back to the origins of medicine itself.

Wild primates from chimpanzees to capuchin monkeys regularly treat themselves with botanicals, choosing specific seeds, leaves, and bark known to have healing properties. When researchers in the Central African Republic observed a gorilla plucking jungle-sop seeds from the dung of elephants, no one was surprised to learn those seeds contained potent alkaloids, and that local healers pre-scribed them (as well as the plant's leaves and bark) as a treatment for everything from sore feet to stomach problems. This pattern re-peats itself throughout the tropics: primates shopping around the apothecary of the rainforest to help rid themselves of parasites, or relieve the pain of injury and disease. Few anthropologists doubt that our own ancestors did the same thing, and in fact a study in the Amazon found that hunter-gatherers used a list of plants that closely mirrored those preferred by monkeys. These ancient habits not only lie at the heart of traditional medicines, they continue to spur the development of new drugs.*

To gauge the importance of seeds in modern medicine, I con-tacted David Newman, an expert on drug development at the Na-tional Institutes of Health. He told me that until the mid-twentieth century, a huge proportion of medications came from plants, many of them from compounds found in seeds. Even today, in an era of synthetics, antibiotics, and gene therapy, nearly 5 percent of all new drugs approved for use in the United States come directly from bo-tanical extractions. In Europe, the number is higher. A recent at-tempt to summarize medicinal research on seeds quickly ran to more than 1,200 pages, with contributions from 300 scientists working in labs around the world. Seed extracts play a role in treatments for everything from Parkinson's disease (vetch and velvet bean) to HIV

---

*Self-medication may be common practice for wild primates, and it prob-ably helped inspire many traditional remedies, but it's not something to trifle with. Like the ricin in castor beans, many compounds from seeds and other plant parts are highly toxic at the wrong dosage.

(blackbean, pokeweed), Alzheimer's disease (calabar bean), hepatitis (milk thistle), varicose veins (horse chestnut), psoriasis (bishop's flower), and cardiac arrest (climbing oleander). Like ricin, many of these compounds serve double duty as both poison and cure, and it turns out that another well-known example happens to come from the seeds of the *almendro* tree.

Fresh from their shells, *almendro* seeds look quite a bit like the almonds that give them their Spanish name, but stretched thin and polished to a dark sheen. The first time I roasted a batch, I immediately noticed the sweet, spicy smell that had brought them to the attention of perfumers in the nineteenth century. Known by the trade name "tonka beans," the fragrant seeds also became popular as a vanilla substitute and as a flavoring for pipe tobacco and spiced rum. Commercial varieties came from an Amazonian *almendro* closely related to the ones I studied in Central America. They spawned an industry that was briefly lucrative enough to warrant huge tonka bean plantations in Nigeria and the West Indies. A French chemist isolated the active ingredient and called it *coumarin* in honor of an Indian name for the tree, *cumarú*. Things went along cheerfully for tonka bean farmers until the 1940s, when researchers discovered that coumarin was toxic to liver cells. Regulators warned that even small amounts could be harmful, and soon banned it altogether as a food additive. Needless to say, consumption of tonka beans has since plummeted, though daring chefs continue to add a few shavings to specialty chocolates, ice creams, and other desserts.

I knew this history when I sat down to sample a few of the roasted seeds with my dissertation adviser, and coauthor of all my *almendro* papers, Steve Brunsfeld. The fact that he was a liver cancer survivor didn't faze us. In botany, tasting weird things goes with the job description—flavors and smells are often valuable tools in identifying plants. Still, we limited ourselves to a few nibbles, just enough to appreciate what struck me as a combination of vanilla and cinnamon, with a citrus finish. Steve wrinkled his mustache and described the flavor more simply: "These things taste like furniture polish." The comment was typical Steve: sharp, funny, and straight to the point. But our *almendro*-tasting moment was also deeply ironic. At the time, neither of us knew that Steve's cancer had reawakened and spread to other parts of his body, and that within a few months his doctors would probably start prescribing a variation of the very compound we'd been joking about.

Since the heyday of tonka beans, scientists have found traces of coumarin in a wide range of plants. It adds to the cinnamon fragrance of cassia bark, and it freshens the smell of cut hay from any field containing vernal grass or sweet clover. But scientists also noticed that something strange happens when plants containing coumarin start to rot. The presence of blue mold and other common fungi changes coumarin from a moderate liver toxin into a blood thinner potent enough to kill a full-grown cow. This discovery solved the riddle of why spoiled fodder sometimes wipes out a farmer's stock. But once researchers mastered that small chemical tweak, it led to billion-dollar advances in two industries: pest control and pharmaceuticals.

Named *warfarin* after the group that funded the research (the Wisconsin Alumni Research Foundation), this modified coumarin quickly became the most widely used rat poison in the world. Mixed in with a tempting food bait, it kills rodents by causing anemia, hemorrhaging, and uncontrollable internal bleeding. But in people, a small dose thins the blood just enough to prevent dangerous clots inside the veins, one of the most common and deadly side effects of cancer and its treatment. Sold under the trade name Coumadin, a warfarin prescription often goes hand in hand with chemotherapy, particularly when the cancer spreads widely, like Steve's. It's also commonly used by stroke and heart patients, and remains one of the world's top-selling drugs over half a century after its discovery.

All the while Steve and I worked on the *almendro* project, his body was fighting cancer. It was a situation that sick botanists must face all the time: struggling against diseases whose treatment may come from the very plants on their herbarium sheets and microscope slides. Steve never told me whether he was taking warfarin, but it wouldn't have been the first time his research had overlapped with the medicine cabinet. He spent much of his career studying willow trees, the original source of aspirin, and once helped a biotechnology company find a good natural supply of false hellebore, a member of the lily family whose toxic seeds, leaves, and roots contain promising anticancer alkaloids.

In the end, no prescription was enough—Steve died a few short weeks before I defended my dissertation. It bothered him to leave things unfinished, in the lab as well as in his personal life, and he kept working long past the point where most people would have thrown in the towel. But while nothing could buy him more time on earth, he did live long enough to get some answers, to know what the research meant. And, for a curious mind like his, that was at least some reward. In the years since, I've often missed not only Steve's friendship, company, and wicked sense of humor, but also his keen intellect. He had that rare ability to cut right through extraneous information, what he would have called "the bullshit," and get at the heart of things. That's a valuable skill both for conversation and for science, because, in nature, even straightforward ideas are rarely as simple as they seem.

On the surface, the notion of lethal seed poisons seems to make perfect sense. It's a natural extension of the same adaptations that led to spices, caffeine, and other defensive compounds. After all, what better way to protect your seeds than to kill anything that tries to eat them? But in practice, taking that evolutionary step from disagreeable to deadly is more complicated. When a seed is attacked, the plant's first imperative is to make the attacker stop, which is why bitterness, pungency, and burning sensations are so common. Immediate physical discomfort drives seed-eaters away and teaches them not to try again, a lesson they can even pass along to others. Poisons, on the other hand, may take hours or days to have an effect, which does nothing to stop a seed attack in progress. A flavorless toxin like ricin makes it theoretically possible for an animal to consume and destroy every seed on a castor bean plant, then wander off and die without even knowing the cause. (And certainly without developing and passing on a "castor-bean-avoidance" behavior!) So while chemicals that cause unpleasantness can discourage whole groups of seed predators, deadly poisons eliminate only individuals, a battle that must be fought again and again. This raises the question of what evolutionary incentive causes some toxins to keep getting stronger, to reach the almost absurd potency of compounds like ricin.

"There don't seem to be any obvious answers," Derek Bewley told me when I posed this question. I hadn't contacted him in a while, but this "god" of seed research was always a generous resource when I ran into puzzles I couldn't solve. He explained that seed poisons often affect different attackers in different ways. Something that evolved to give one animal a modest stomachache (and teach it not to eat that seed again) might prove utterly deadly to another. Or a poison that took days to dispatch large creatures might kill insects in seconds, thereby stopping an attack just as quickly as a foul taste. "Or, the whole thing might be a fortuitous accident," he mused, and pointed again to the castor bean example. "Ricin is an easily and early mobilized storage protein, and its toxic properties might be just a useful side-effect."

When Noelle Machnicki investigated the capsaicin in chili peppers, she learned that what began as an antifungal compound ended up influencing everything from insects and birds to the taste buds of mammals, including people. The same complexity applies to seed poisons, and it would probably take a determined doctoral student like Noelle to unravel the story behind any one of them. But there is something certain about all poisonous seeds: no matter how toxic they may have become, the plant must have also invented some way to disperse them. Because there's no use in keeping your seeds safe if you can't move them around. In the case of castor beans, the solution is twofold: an explosive pod that hurls the ripe seeds up to thirty-five feet (eleven meters) away from the mother plant, and a nutritious little package attached to the *outside* of the seed coat, which makes the seeds attractive to ants. Anywhere in the world, the scene near mature castor bean plants is pretty much the same— pods popping open, seeds flying, and thousands of ants busily dragging them home to their underground nests. Once there, they chew off the food packet and leave the seed untouched, safely buried and ready to sprout. Surprisingly, no one has yet looked into whether the food packets are harmless, or if the ants have developed an immunity to ricin. Either way, this clever system allows castor beans to become exceedingly deadly with no risk of compromising their ability to disperse. For *almendro*, on the other hand, the presence of coumarin in seeds is a little harder to explain.

Though it's not technically rat poison without modification by mold or chemistry, coumarin still seems an unlikely compound for a seed dispersed by rodents. Even in its unadulterated form, it wreaks havoc on their livers. The toxicity that got it banned as a food additive was first noticed in an experiment on laboratory rats. Fed a diet supplemented with coumarin, the rats systematically lost weight, developed liver tumors, and died young. No one has studied this dynamic in the wild, but it's hard to imagine a diet more rich with coumarin than that of agoutis, squirrels, and spiny rats living beneath an *almendro* tree. Yet these rodents continue feasting on the seeds—and occasionally dispersing them—with no apparent ill effects. Have they developed an immunity? Do their livers recover during other seasons, when *almendro* seeds aren't available? Or perhaps they are, indeed, dying young, unnoticed in their nests and burrows. Nobody knows the answer, but there is another possibility that is even more intriguing.

Coumarin occurs in a wide variety of plants, but nowhere is the concentration higher than in *almendro* seeds. (That's why European perfumers continue to get theirs from tonka beans rather than trying to squeeze it out of the vernal grass in their backyards.) Is it possible that the coumarin in *almendro* is on the rise? Could we be witnessing the early stages of a new chemical defense strategy? At the moment, rodents do indeed disperse *almendro* seeds, but that situation is only a snapshot in evolutionary time. From the plant's perspective, it's a messy and chancy business. Agoutis and squirrels will eat and destroy every seed they can, only dispersing the ones they happen to forget about. If *almendro* does develop a coumarin potency strong enough to keep them away, it wouldn't be the first time a seed defense targeted rodents. Remember that capsaicin, to name but one example, burns the mouths of seed-eating rats and mice, but has no effect on the beaks of the seed-dispersing birds. But the *almendro* could only afford to deter rodents if, like chili peppers, it had an ace in the hole, another option for dispersing its seeds. And after walking hundreds of transects in the jungle, and analyzing thousands of samples in the lab, we realized that is exactly what is going on.

Seeds Travel

Each pregnant Oak ten thousand acorns forms
Profusely scatter'd by autumnal storms;
Ten thousand seeds each pregnant poppy sheds
Profusely scatter'd from its waving heads.

—Erasmus Darwin,
    The Temple of Nature (1803)

CHAPTER TWELVE

Irresistible Flesh

Did Nature have in view our delectation when she made the apple, the peach, the plum, the cherry? Undoubtedly; but only as a means to her own private ends. What a bribe or a wage is the pulp of these delicacies to all creatures to come and sow their seed! And Nature has taken care to make the seed indigestible, so that, though the fruit be eaten, the germ is not, but only planted.

—John Burroughs, *Birds and Poets* (1877)

"**M**urciélago," José breathed. "A bat!" And for the first time in our long work together I saw his cool reserve give way to something like wonder. The *almendro* seeds lay on the ground before us, grouped together in a loose pile. Where we usually felt lucky to find one or two, this trove held more than thirty—a veritable mother lode. And yet we knew there wasn't a mature *almendro* tree within half a mile (eight hundred meters), too far for any rodent to carry such a horde. I knelt down and we began collecting the seeds, placing each one in a carefully numbered plastic bag. They were still fresh, their hard shells surrounded by a thin green pulp chewed into damp strands. Looking up, I already knew what I would see—the drooping, twelve-foot (four-meter) frond of a young palm tree, favorite perch of the largest fruit bat in Central America.

With a wingspan stretching to eighteen inches (forty-five centi-meters), the great fruit-eating bat has more than enough flapping power to carry off *almendro* seeds. In flight, those huge wings make its four-inch (ten-centimeter) body look like an afterthought—just enough bones and skin to hold on to a heavy load. While fruit-eating bats often dine on figs, flowers, or pollen, the pile before us proved that *almendro* was something special—for the bat, and for the tree itself. Unlike squirrels and agoutis, who eat and destroy any seed they don't misplace, the fruit-eating bat's interest is entirely eponymous. It wants only the thin, watery flesh that surrounds the shell. Perched upside down, with sharp teeth working furiously, a bat can scrape away the pulp from the husk in minutes, dropping the seed unharmed to the forest floor below.

To me, eating *almendro* fruit was like gnawing on a bland, over-ripe snap pea, toughened by sun. But for the bat (or bats) that had roosted here, it was a taste sensation worth thirty round-trip flights. And worth the risk of returning again and again to a tree where owls, bat falcons, and pythons lurked, waiting specifically for an opportunity to snatch the unwary visitor. That element of danger played a vital role in the system. Without it, the bats would simply lounge in the *almendro* itself, gorging on fruit and dropping the seeds directly below the mother tree—unharmed, but undispersed. (Mon-keys do exactly that during the daylight hours, as do smaller bats incapable of lifting the heavy fruits.) But so long as predators staked out the trees, any bat large enough would carry its prize away to the safety of a feeding roost, creating a pattern of seed dispersal so dis-tinctive that José and I came to know these bats' every move, with-out ever laying eyes on one.

I glanced again at the empty palm leaf above us before we walked on. It was a familiar sight. To confirm our hunch, we had looked up in exactly the same way nearly 2,000 times, comparing the locations of palm leaves to the locations of seeds. Whether we found them sin-gly, in pairs, or in troves like this one, our largest yet, the dispersed progeny of *almendro* were twice as likely to lie beneath a bat roost.

(The bats chose palm fronds for good reason: the drooping leaflets hid them from predators above, while the long, spindly stalk would shake with warning if anything tried climbing up from below.) This pattern held everywhere we looked—in isolated patches as well as large swathes of virgin forest. Back in the lab, genetic fingerprinting helped me take the data further. By tracking particular seeds from tree to roost, I could show that a bat would fly almost anyplace that an *almendro* came into fruit. Even trees stuck in the middle of pastures were part of the network, attracting hungry bats and co-opting them into carrying their seeds to better habitat thousands of feet away. With large rainforests disappearing, our results gave me hope that *almendro*—and the many species that depended on it—could persist in this new landscape of fragments, farms, and pastures.

Hiking back along our transect, we passed suddenly into blinding sunlight where the forest ended in a straight green line. Rank grassland stretched away over rolling hills dotted with remnant trees, some of them *almendro*. We knew the area well and weren't surprised to see the landowner, Don Marcus Pineda, leading a donkey across a nearby field. He waved and turned in our direction. Pineda owned a lot of property and still worked it himself, clearing trees, mending fence lines, and tending a large herd of beef cattle. As he neared, I smelled something chemical from the sloshing yellow jugs strapped to the donkey's packsaddle. Pineda told us he was heading out to spray some bracken, a prolific, inedible fern he wanted gone from his rangeland. But we knew there was more news, or he wouldn't have gone so far out of his way to greet us. Finally, he spoke again.

"*El Papa ha muerto,*" he said simply—"The Pope is dead." Marcus Pineda lived on a rugged frontier farm near the Nicaraguan border, and had always struck me as machismo personified—a tough, lined face squinting out from beneath his ever-present cowboy hat. But it was obvious this loss had hit him hard, and it shook José, too. For several minutes the three of us stood together quietly, heads bowed in the muggy heat. Pope John Paul II had been a hero in Costa Rica, where over 70 percent of the population identified as Roman

Catholic. But he was more than a religious leader. His frequent trips to Latin America, his personal charisma, and his genuine interest in the region made him a beloved figure both inside and outside the church.

As a scientist, I, too, felt a fondness for John Paul. After all, he was the pope who finally pardoned Galileo, and he did more than any of his predecessors to reconcile church teachings with the theory of evolution. In discourses to the Pontifical Academy of Science, he had called Darwin's ideas "more than a hypothesis," and had gone so far as to imply that the Book of Genesis was allegorical, and not "a scientific treatise." His words to the academy were brief, but if John Paul had spoken at length, he might have pointed out any number of metaphors in Genesis, many of them biological. The chapters concerning Adam and Eve, for example, do more than describe the dawn of humanity and original sin. They also tell one of the greatest seed dispersal stories of all time.

From the Renaissance forward, artists have made the scene indelible: Adam and Eve sharing a luscious apple below the Tree of Knowledge of Good and Evil, with a serpent coiled around the closest branch. Botanical purists point out that such large-fruited apple varieties didn't become common until the twelfth century, and that the fruit should probably be a pomegranate. Whichever the species, the cunning snake had chosen a perfect lure, something that evolved for the sole purpose of temptation. To a hungry animal, the tiny pips inside an apple or the stone at the center of a date may seem irrelevant, secondary to the irresistible flesh. But the truth is the other way around. Fruit, in all its magnificent variety, exists for no other reason than to serve the seeds.

Whether a plant is growing in the Garden of Eden, in a tropical rainforest, or in a vacant lot, its investment in producing, nourishing, and protecting its seeds means nothing without dispersal. Offspring that languish on the mother or drop directly below amount to little more than a wasted effort. If they sprout at all, they won't survive long in the shade of a fully grown parent. (In some cases,

FIGURE 12.1. Albrecht Dürer's 1504 engraving of
Adam and Eve has it all—fig leaves, a tree, a snake,
and the ultimate sign of temptation: fruit. *Adam and
Eve*, Albrecht Dürer, 1504. WIKIMEDIA COMMONS.

adults release toxins into nearby soil to prevent their progeny from
becoming competitors.) For *almendro*, adding a thin layer of pulp to
its seeds can entice fruit bats to carry them half a mile or more. The
Tree of Knowledge did even better. According to Genesis, eating
that Forbidden Fruit resulted in Adam and Eve's immediate expul-
sion from Eden. Metaphorically, at least, the fruit went with them.
Some depictions show the guilty couple still clutching a half-eaten
apple. And if it was indeed a pomegranate, then the seeds would
have been safely lodged in their digestive tracts. Either way, the Tree
had put itself in a great position. With that one tempting fruit, it

went from a garden-bound existence to the promise of mass dispersal with humanity across the face of the earth.

Much has been written about the relationship between people and fruit or other crops—the way we take them with us wherever we go. Apples alone went from a single species domesticated in the mountains of Kazakhstan to thousands of varieties—people grow them on every continent outside Antarctica. It's only a slight exaggeration to call us servants of our food plants, diligently moving them around the world and slavishly tending them in manicured orchards and fields. And it's no exaggeration at all to call this activity seed dispersal. We do it as unconsciously as the bats do, living out an interaction between plants and animals that is nearly as ancient as seeds themselves. Fruit influences our behavior because it evolved to do so; developing flesh we find sweet and colors and shapes that attract our attention. Its power reaches beyond our farms and kitchens, touching beliefs at the boundary between culture and imagination. Look no further than the glut of grapes, pears, peaches, quince, melons, oranges, and berries festooning every basket and platter in the history of still-life art. Our desire for fruit makes it more than a symbol of temptation—it helps us to define beauty itself.

In nature, fruit is typically both delicious and fleeting, traits that help it attract just the right dispersers at just the right moment. People generally seek out sweetness, but plants can readily develop fruits to please other palates, producing proteins and fats as well as sugars. The rich packets adorning castor beans (and a host of other species) are designed to attract otherwise carnivorous ants, while the Kalahari Desert's *tsamma* melon, ancestor to the watermelon, draws in all comers by satisfying a universal hot-country yearning: thirst. In any case, the desired flavor appears only as the seeds mature and prepare for departure. Before the seeds ripen, plants keep animals at bay with fruit that is bitter or downright poisonous. The physician who accompanied Christopher Columbus on his second voyage observed a group of sailors on the beach happily tucking into what appeared to be wild crabapples. "But no sooner did they taste

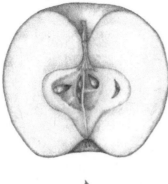

FIGURE 12.2. Apple (*Malus domestica*). An iconic symbol
of temptation in everything from artwork to Bible stories
to *Snow White*, apples play a role uniquely suited to fruit.
In nature, fleshy fruits of all kinds evolved for the sole
purpose of tempting animals into dispersing the seeds of
plants. ILLUSTRATION © 2014 BY SUZANNE OLIVE.

them than their faces swelled, growing so inflamed and painful that
they almost went out of their minds." The men survived, but had
probably been trying to eat *manzanillo*, a fruit the local Carib Indians harvested for making arrow poison. It remains toxic even when
ripe, perhaps as a deterrent to insects or fungi, or to ward off all but
a specialized (and as yet unknown) seed disperser. Poisonous, single-
species strategies are unusual, however. Most fleshy-fruited species
take the route of the apple, luring potential dispersers with something as widely desirable as the plant can afford to produce.

"Affordability" may not sound like a botanical term, but bal-
ancing the household budget dominates the lives of plants. Energy,
nutrients, and water are the coins of the realm, limited resources
that must be divided among vital priorities. Spending a fortune on
dispersal runs the risk of shortchanging the seeds' nourishment or
protection, not to mention the growth and defense of leaves, stems,

and roots. In the world of plant economics, producing fleshy fruit is costly. Gardeners and farmers know this from experience—the "heavy feeders" in any vegetable plot always include the large-fruited crops like tomatoes, melons, squashes, eggplants, cucumbers, and peppers. Adding fertilizer or a scoop of compost tilts the balance, helping these species invest more in the production of something succulent. In the wild, plants scrape by with whatever the local soil and weather provide. But even in a good year, the huge expenditure of fruiting almost always makes the season brief, only adding to its cachet.

As one of the rarest, sweetest, and most nutritious things in the landscape, ripe fruit can draw animals from far and wide. African elephants trek miles out of their way to find favorite species like bitterbark, an odorous, cherry-sized fruit native to the Congo basin, or *marula*, a southern African delicacy related to mangos. In one forest, researchers mapped a network of elephant trails connecting every known adult of the *balanites* tree, even though it only bore fruit every two or three years. Our own species also goes to great lengths to take advantage of a wild harvest. Traditional San tribespeople in the Kalahari base their travel routes and seasonal encampments around the availability of *tsamma* melons, just as Australian Aboriginals in the Western Desert once did with figs, wild tomatoes, and *quandong*, a peach-like member of the sandalwood family. It's not uncommon for similar fruit interests to put people and wildlife in direct competition. Friction between villagers and mountain gorillas in Uganda increases during *omwifa* season, when both apes and local residents home in on the same wild groves to harvest the tree's lumpy, aromatic fruits.

The allure of fruit begins with biology but endures in countless cultural references, from Chinese symbols for immortality (peach), wealth (grape), and fertility (pomegranate) to the traditional American token of welcome (a pineapple). Strawberries fed Freyja, the Norse goddess of love, while the Greeks paid tribute to Athena as the inventor of olives. In Southeast Asia, the Hindu deity Ganesh was famously fond of mangos, and the spreading boughs of the bodhi fig shaded both the birth of Vishnu and the enlightenment of the Buddha. (It's a species so sacred that even taxonomists got the message, giving it the botanical name *Ficus religiosa*.) The bounty of the biblical Eden reflects a long tradition of describing Paradise as a decidedly fruit-laden place. In the words of the poet Hesiod, those fortunate enough to reach Greece's famed Elysian Fields enjoyed "honey-sweet fruit flourishing thrice a year." Islamic texts allude to eternal gardens filled with everything from dates, cucumbers, and watermelon to "the quince of Paradise." Medieval Britons made matters even more plain, referring to the mythical homeland of King Arthur as *Avalon*, from the Welsh for "Island of Apples." Scholars of etymology trace the very word *Paradise* to a Persian term for a walled enclosure that early Hebrews adopted to mean "fruit garden," or simply "orchard." But perhaps the best example of our esteem for fruit comes from English, where any successful venture is considered *fruitful*, and failures are known as *fruitless*.

With fruit so powerfully embedded in language and culture, it's easy to forget that from a functional standpoint, sweet flesh is simply window dressing for seeds—an elaborate means of travel from here to there. The technical term for fruit dispersal is *endozoochory*, which would sound far more elegant if everyone still spoke ancient Greek: "going abroad within animals." (We scientists have a great fondness for long mash-ups in dead languages. When a bat moves an *almendro* seed, for example, the proper description is *chiropterochory*: "going abroad with an animal whose hands resemble wings.") This form of exploitation evolved early in the history of seed dispersal, almost as soon as there were creatures large enough to get the job done. Back in the Carboniferous Period, the forests that Bill DiMichele now studies on the ceilings of coal mines harbored a relatively modest community of insects, amphibians, and early reptiles. But seed ferns and primitive conifers soon pioneered a range of strategies to ensnare them, from small, skinny seed packets, which may have attracted millipedes, to fleshy seeds the size of mangos, which probably stank like rotten meat, a siren call to ancestral dinosaurs. Echoes of that era remain in the pungent seeds of ancient survivors like ginkgos, whose reek is so strong that many cities ban the planting of female trees. Virtually every ornamental ginkgo in the world is a male, producing nothing more offensive than odorless pollen.

Botanically speaking, early seed plants lacked the specific tissues that would qualify as a true "fruit." But that didn't stop them from developing analogs that worked just as well—sweetening the outer layer of a seed coat, for instance, or putting flesh on a nearby stem or bract. Modern conifers and other gymnosperms continue that tradition, as gin drinkers know well from the pulpy, aromatic cones known as juniper "berries." But while animal dispersal remains common in gymnosperms, most familiar fruits evolved with the angiosperms—the flowering plants—whose seeds came in a package, by definition. That covering opened up a world of fruity possibilities, and it arose right alongside an explosion in the availability of dispersers. Birds, mammals, and flowering plants all experienced what taxonomists call a *radiation*, a rapid increase in species, immediately following the extinction of the dinosaurs. And while older groups, like lizards, insects, and even fish, continue to disperse seeds, the vast majority of fleshy fruits are meant to attract—and go abroad with—birds and mammals.

The best way to appreciate fruit diversity involves a simple experiment that most of us undertake several times a week: shopping for groceries. Even a tropical rainforest can't compare to the density of species arrayed in a typical supermarket produce aisle. My hometown grocery store has been on the same block since 1929, strategically located between two other longtime businesses—the drugstore and the local tavern. On a recent spring morning, the shelves boasted seventy-one varieties of fresh fruit from thirty-nine distinct species. The smallest was a blueberry no larger than my thumbnail. Produce departments now stock them year round, but in the wilds of North America, where blueberries evolved, their ripening coincides with fall bird migrations and the pre-hibernation fruit-gorging

of bears. At the other end of the spectrum, I found a bin of watermelons that weighed up to fifteen pounds (seven kilograms) each.
Their *tsamma* melon ancestors ripen during the dry season in southern Africa, providing essential water for everything from antelopes
to hyenas to people. Every fruit in the store told some version of the
same story—a wild scenario made commonplace by the efficiencies
of modern agriculture. Of course, many tree fruits are now propa-
gated by cuttings, and most of the seeds in the store would eventually end up in someone's compost bin or septic system, but their
mere presence demonstrated the success of the fruiting strategy.
The everyday bonanza of grocery-store produce is something like
extreme dispersal in action—the fruits on display had come from as
far away as Italy, Chile, and New Zealand. But they were not only
well traveled. They also served as a lesson on the range of ways that
flowering plants make fruit, from the obvious—like the sweet flesh
of an apple—to the ones that most people haven't really thought
about—like the juice-filled hairs inside an orange, or the strawberry,
whose shape and flavor come from a swollen flower base, which is
why the seeds perch so oddly on the *outside* of the fruit.

The interplay between fruits and their dispersers affects every
partner in the dance. It influences dietary habits and migration patterns as well as the timing of reproduction—for both the animals
and the plants involved. But the adaptations can be much more
specific. The teeth of fruit bats, for example, evolved from insect-
eating cleavers into a bite with angular surfaces designed to crush
and pulverize. Guenons and vervet monkeys have special pouches
that stretch from their cheeks down the sides of their necks, allowing them to stuff in huge loads of fruit for safe consumption later.
Startle one of these monkeys at a fruiting tree, and your last sight
of it will be a distinctly bulging face leaping away through the canopy. Fruit-eating birds have developed everything from wider beaks
and flexible throats to shorter intestines for processing abundant
fruit supplies quickly. Parrots counter the toxins in unripe fruits
by gobbling up clay rich in *kaolinite*, the same mineral that formed

the original basis for the stomach-soothing tonic Kaopectate. I've watched Cedar Waxwings eat clay, too, which is hardly their only fruity distinction. They digest berries so quickly that their droppings are still sweet (which led them to develop unique rectums that absorb sugars just as well as their intestines do).

Plants, in turn, have learned to tailor their strategies to attract particular types of dispersers. Birds appreciate prominent splashes of red or black (raspberry, blackberry, cranberry, black currant, hawthorn, holly, or yew), but odor is more important for drawing in animals that are color-blind (elephants), nocturnal (bats), or whose noses are sharper than their eyes (tortoises, opossums). The pits and pips at the centers of fruit boast some of the hardest seed coats in nature, tough enough to withstand scraping, chewing, and the chemical scour of digestion. In fact, being eaten by a disperser enhances the germination of fruit seeds twice as often as reducing it. When an elephant chews the fragrant fruit of a South African marula tree, its tremendous teeth loosen woody plugs in the pit, an essential step that later allows the seed to imbibe water and sprout. The exact benefit isn't always so clear, but digestion boosts the germination of everything from the cherries devoured by bears to the prickly-pear cactus preferred by the Galapagos tortoise. Some combination of chemical change and physical abrasion probably helps to break the seeds' dormancy, and then there's the end result: depositing those seeds in a warm pile of fertilizing dung. In some cases, other creatures then gather up the seeds and disperse them farther. Tree squirrels do it with the marula nuts from elephant piles, and deer mice scatterhoard the chokecherries and dogwood seeds they find in bear poop. But the most notorious example of this process takes us back to the world of gourmet coffee, where people pay up to $300 per pound ($650 per kilogram) for coffee beans plucked from the turds of the Asian palm civet. A single cup in a trendy Manhattan café can set you back nearly $100. That high price tag has inspired lucrative spin-offs from the poop of other species known to nosh coffee berries: elephants in Thailand, Peruvian coatis, and a turkey-like Brazilian bird called the Dusky-legged Guan. (Unfortunately, the civet coffee boom has also resulted in cruel schemes to force-feed caged animals. When the topic came up during my visit to Slate Coffee Bar in Seattle, barista Brandon Paul Weaver dismissed the whole craze with a memorable line: "Coffee from assholes, for assholes.")

Dispersal by way of tasty flesh occurs in nearly a third of all plant families, from cycads to squashes to citrus. The strategy evolved again and again in different settings, because when it works, the results are dramatic. A thirsty brown hyena, for example, can eat eighteen *tsamma* melons in a single night, dropping their seeds over a home range as large as 150 square miles (400 square kilometers). Brown bears in a blueberry patch do even better, munching their way through 16,000 of the tiny fruits in a matter of hours. Since each berry contains an average of thirty-three seeds, that puts the blueberry dispersal rate of a single hungry bear at more than half a million seeds per day. Examples abound, and whole scientific careers have been happily devoted to exploring the nuances. But the fact remains that most seeds travel by other means.

For the vast majority of plants, seed dispersal represents the one moment of mobility in an otherwise stationary life. It determines what grows where, a fundamental organizing principle of ecosystems. As such, the process carries great evolutionary consequence, and seed plants have been coming up with variations on the theme for close to 400 million years. Fruit offers a reward, but other seeds simply hitchhike, using hooks, spines, or stickiness to catch a free ride on the *exterior* of an animal. (The commercial form of this strategy is known as Velcro, a product inspired by the way burdock seeds stuck to the fur of the inventor's dog.) Some seeds launch from exploding pods, while others drop into water and drift with the tide. For many people, the most familiar dispersal experience comes from the poke of grass seeds lodged in a sock. They work their way inward with every step, eventually becoming so unbearable you're forced to stop, pluck them out, and throw them on the ground. Mission accomplished.

In Costa Rica, José and I learned that *almendros* move around with help from the wings of bats. But long before bats even existed, seeds had learned to grow wings of their own. Whether gliding, twirling, wafting, or soaring, riding the wind is the most ancient form of seed dispersal, and it remains the most common. With all that practice, plants have developed means of flight (and in some cases, flotation) that transport seeds in quantities, and over distances, never dreamed of by bats, bears, or birds. The results do more than organize the placement of shrub, herb, grass, and tree. They give us yet another chapter in the long story of seeds and people—how papery wings and bits of fluff influenced everything from aeronautics and fashion to the history of industry, the British Empire, and the American Civil War. And like so many biological tales, the best place to start lies in the notebooks and journals of a certain young naturalist in the Galapagos Islands.

CHAPTER THIRTEEN

By Wind and Wave

*The vegetable life does not content itself with casting from the
flower or the tree a single seed, but it fills the air and earth with
a prodigality of seeds, that, if thousands perish, thousands may
plant themselves, that hundreds may come up, that tens may
live to maturity, that, at least, one may replace the parent.*

—Ralph Waldo Emerson,
Essays: Second Series (1844)

As a young man, Charles Darwin didn't much care for plants.
When he set forth as naturalist on the HMS *Beagle*, botany
trailed a distant third behind his passions for geology and zoology.
He described himself as "a man who hardly knows a Daisy from a
Dandelion," and even his mentor at Cambridge, who recommended
him for the job, admitted, "he is no botanist." Later in life, botanical studies came to dominate Darwin's research, and he wrote
whole books about carnivorous plants, climbing plants, the structure
of flowers, and the pollination of orchids. But as the *Beagle* slowly
wended its way around the coast of South America, Darwin pursued
his botanical collections mostly as a matter of duty, and even considered throwing the specimens away. So when the ship finally made
landfall in the Galapagos, it's not surprising that he devoted most of

his attention to the volcanoes, lava fields, tortoises, and odd birds. A single passing comment in his field notebook seemed to sum up Darwin's thoughts on the flora: "Brazil without big trees." Of his first day ashore, on Chatham Island, he later wrote: "Although I diligently tried to collect as many plants as possible, I succeeded in getting only ten kinds; and such wretched-looking little weeds would have better become an arctic, than an equatorial Flora."

But while Darwin may have been more excited about the craters and finches, his botanical diligence in the Galapagos paid off handsomely. Over the next five weeks he managed to gather and preserve 173 species, nearly a quarter of the known flora. And in the years ahead, those plants would add a critical dimension to his ideas about evolution. Because Darwin's thoughts on how species arose had in fact started with a question anyone might ask on a journey as long and far-flung as the *Beagle*'s: Why do plants and animals occur where they do? Scholars still argue about how much the Galapagos influenced Darwin's thinking, but as early as his fifth day in the islands he jotted down a revealing note: "I certainly recognise S. America in ornithology, would a botanist?" Clearly, he was already wondering where the ancestors of the Galapagos flora had come from. As it turns out, among those wretched-looking weeds from Chatham Island was a plant capable of answering that question perfectly. When his friend Joseph Hooker examined the specimen a few years later, he immediately noticed both similarities to, and differences from, its South American cousin. Botanists now call it *Gossypium darwinii*—Darwin's cotton—and the story of how it reached the Galapagos is a powerful example of just how far seeds can travel, why they do it, and what happens when they get there.

The scientists who study cotton didn't have to invent a Latin name for the genus; they simply adopted "*Gossypium*" straight from the Romans. Cotton was a well-known fabric throughout the ancient world. The armies of Alexander the Great brought the first examples back from India, and it soon spread around the Mediterranean and south to the Arabian Peninsula (where people called it

FIGURE 13.1. Cotton (*Gossypium* spp.). Lined up end to end, the fibers from a single cotton boll can stretch more than twenty miles. Woven together into yarn, they anchor an industry that shaped the history of empires, the Industrial Revolution, and the American Civil War. A full boll is pictured above, with fuzzed and shorn seeds shown below. ILLUSTRATION © 2014 BY SUZANNE OLIVE.

qutun, the source of its name in English). The Aztecs and Incas had cotton, too, as did the Arawak Indians encountered by Christopher Columbus. Wherever he went ashore in the Caribbean, he found people weaving it into everything from fishing nets and hammocks to women's skirts that he described as, "just big enough to cover their nature but nothing else." Columbus mentioned Arawak cotton nineteen times in the log from his first voyage, noting, "They do not plant it by hand, for it grows naturally in the fields like roses."

That same observation could have been made in tropical places around the world, where more than forty different species of cotton grow wild. Some have simple seeds, but wherever the pips are plumed, local people have learned to spin those fibers into thread. Cotton now reigns as the most popular fabric in the world, anchoring a $425 billion industry that makes it the most valuable nonfood crop in history. It's so ubiquitous we have to be reminded that it didn't evolve to be woven into togas, turbans, hammocks, and t-shirts. The elaborate fluff engulfing a cotton seed arose with a different purpose in mind—helping baby plants ride the wind.

To understand the concept of wind dispersal, simply let your lawn grow long and weedy in the springtime, and take a walk there with the nearest child. Since his first days on two legs, our son Noah has been fond of plucking ripe dandelion heads and holding them upward with an irresistible, one-word request: "Blow!" By my count, one modest puff can send over two hundred seeds suddenly airborne, drifting down in a delight of tiny parachutes. Try to catch them, and you'll soon find yourself running five, ten, or even twenty feet (six meters) from the mother plant. On a windy day, they will escape you completely. With dandelions now common from London to Tokyo to Cape Town, this ritual has become a universal lesson in botanical aerodynamics, the architectural upshot of seed and breeze. For dandelions, the trick lies in a delicate spindle tufted with lint—symmetrical, flexible, and perfectly spaced for maximum drift. Cotton stays aloft with a different design, and to understand it I decided to do something that few people have bothered with since Eli Whitney invented his famous cotton gin: pick apart a cotton boll by hand.

In the wild, cotton plants grow as perennial shrubs or small trees, with angular branches and hairy, gray-green leaves. Domestic varieties are fast-growing annuals—shorter in stature, but otherwise much the same. They all belong to the mallow family, a large group that includes Congo jute and okra, but is best known for showy garden flowers like hollyhocks and hibiscus. Cotton has beautiful flowers, too, with papery, lemon yellow petals surrounding a purple center. The fruit forms a round pod, or boll, that bursts and inverts as it ripens, revealing plumes of white down that make cotton fields look

FIGURE 13.2. Stories from medieval traveler Sir John Mandeville and others led to the myth that cotton came from "vegetable lambs," woolly botanical creatures harvested from the fruits of an Asian tree. Anonymous (c. seventeenth century). WIKIMEDIA COMMONS.

like a harvest of snowballs or fantastical sheep. In the fourteenth century, English traveler Sir John Mandeville astonished readers back home with descriptions of an Asian tree that bore gourd-like fruits bursting with tiny lambs. It's unclear whether he meant cotton, which he described more accurately elsewhere, but the idea took hold. Embellished versions of the story soon attributed cotton to "vegetable lambs," and illustrators pictured them stretching their fuzzy necks down from the branch tips to graze.

Like Mandeville, I come from a cool, rainy island where the idea of growing cotton sounds extremely exotic. Unlike a medieval Englishman, however, I didn't have to travel to India to find some in its natural state. Modern craft stores offer raw cotton bolls at a very reasonable price, still attached to the branch. They're meant for wreaths and flower arrangements, but each one holds an incredible tale of seed evolution for any plant adventurer brave enough to attempt a dissection. Armed with tweezers, a pocketknife, and a pair of sharp microscope probes, I clipped a medium-sized boll from the branch and headed for the Raccoon Shack.

Stem side down on my desk, the boll did indeed resemble a sheep, its tufted white back as soft as old flannel. But when I pressed the fuzz tight between my fingers I felt the nubs of seeds deep inside, completely buried in fibers. The boll measured three inches by two (seven and a half by five centimeters) and weighed an eighth of an ounce (four grams), a size oddly similar to the small wren I had taken apart on the same desktop while researching my book on feathers. It, too, had been light, compact, and designed to fly. Plucking the bird involved two solid hours of painstaking tweezer-work—gripping, tugging, and sorting through more than 1,200 tiny plumes. That was child's play. After less than a minute of hand-ginning cotton, I realized I didn't stand a chance of disentangling even a single lone fiber. They weaved and snaked together so tightly I couldn't pull on one without yanking knots in scores, if not hundreds, of others. My plan to neatly tease apart seed from felt collapsed. I had hoped not only to gin the cotton, but also to count, sort, and measure individual fibers, just as I had done with wren feathers. In the end, I resorted to scissors, and even then it took dozens of hacking snips to cut through the woolly mat. The final result was a snow-drift of tangled cotton and a pile of pathetic seeds, still furred with uneven clumps of fiber. The analogy of a poorly shorn flock sprang inescapably to mind.

Under a microscope, the source of my problems became clear. At the surface of each seed, dense fuzz sprouted like close-cropped turf, so thick I couldn't make out where the seed coat ended and where the fluff began. A glance at the cotton entry in Derek Bewley's seed encyclopedia told me why: they were one and the same. When it comes to seed coats, plants like cotton don't follow the rules. Instead of protection, a cotton seed devotes its outermost layer to the promise of dispersal. Flight (and flotation, as we will soon learn) has provided an evolutionary incentive great enough to transform individual cells from microscopic blips into epic filaments over two inches (five centimeters) long. It's no wonder they're hard to untangle. With every hair only one cell wide, cotton seeds the size of a split pea can easily grow a downy coat of more than 20,000 fibers. With thirty-two seeds in the average fruit, that makes every boll a tangle of more than half a million strands. Lined up end to end, they would stretch more than twenty miles (thirty-two kilometers).

It's said that Eli Whitney took inspiration for his famous machine from the sight of a barnyard cat pouncing on a chicken. The bird squawked and scooted off, leaving the cat with tufts of feathers clinging to each claw. Cotton gins work the same way, clawing fiber from seed with hooks fixed to great spinning drums. Whitney's 1793 patent application pictured a humble wooden box with a hand-crank and a single roller. That technology quickly advanced through the ages of steam and electricity, all the way up to modern computerized behemoths that can sort, clean, dry, and press a 500-pound (227 kilogram) bale in less than two minutes. Through it all, the evolutionary intent of cotton remained abundantly clear; drifting and spinning in wispy clouds around every gin in what one nineteenth-century observer called a "furious snowstorm." The single-celled hairs on cotton seeds combine maximum surface area with minimum—indeed barely perceptible—weight. Whether sent aloft by wind or machine, the fluff stays airborne and moves around exactly as the seed intended.

Dispersal by wind leads to what biologists call a "seed shadow." Breezes are fickle, and even the best aerodynamics can only keep something flying for so long. The result is a predictable pattern: most seeds drop to earth fairly close to the mother plant, falling in a dense cluster that thins to scattered, then occasional, and then rare the farther away you look. Just where that seed shadow ends remains an open question, since truly long-distance events occur too infrequently to study. One of the few attempts to track seeds in the atmosphere focused on a weedy aster called Canadian fleabane. Using a remote-controlled airplane equipped with sticky seed traps, researchers found fleabane plumes ascending on updrafts to at least 375 feet (120 meters). From there, even a modest wind can move them tens or hundreds of miles or kilometers. But we know that seeds can waft much higher and farther. In the Himalaya, unidentified wind-blown seeds have been found in rocky crevices at 22,000 feet (6,700 meters), far above the realm where plants can grow. No one knows how far they've traveled, but enough of them arrive to form the basis of a food chain: fungi rot the seeds, springtails eat the fungi, and tiny spiders prey upon the springtails.

The best evidence for long-distance dispersal, however, does not come from mountaintops or from combing the upper atmosphere for the occasional seed. It's found in the very patterns that so fascinated Charles Darwin during his journey on the *Beagle*—the distribution of species—and in the simple fact that plants growing in remote places, like cotton on the Galapagos Islands, couldn't have gotten there by any other means. Darwin's notes don't reveal exactly when he started thinking about seed dispersal, but within a few years of the *Beagle*'s return to England, he began immersing everything from celery seeds to whole asparagus plants in flasks and tanks of seawater. Most species germinated well after a month, and some lasted longer than four, but he was disappointed when only a few seeds managed to stay afloat past the first few days. Still, he calculated dispersal distances of at least 300 miles (483 kilometers) on an average Atlantic current, and he pondered the idea of wind-dispersed seeds floating up on distant beaches, drying out, and tumbling inland on the breeze.

Darwin's experiments focused on the plants commonly found in an English garden—cabbage, carrots, poppies, potatoes, and so on. From this humble beginning he cautiously concluded that long-distance dispersal by ocean current, as well as by wind and birds, could help explain the colonization of islands like the Galapagos. But he still harbored doubts about how far seeds might travel, and their fate once they arrived: "How small would the chance be of a seed falling on favorable soil, and coming to maturity!" Had he tried his experiments on cotton, he might have felt more encouraged.

It turns out that the same fluffiness that keeps cotton aloft in wind also helps it to float in water, trapping air bubbles that make the seeds of long-fibered species buoyant for at least two and a half months. Their dense hairs also keep water from penetrating the seed coat—even after cotton seeds sink, they can remain viable in salt water for more than three years. With genetic data now matching Darwin's cotton to a coastal South American ancestor, researchers have a pretty clear idea of just how it crossed the 575 miles (926 kilometers) from mainland to archipelago. Blown far out to sea by a storm, or what is known in the dispersal business as "an extreme meteorological event," that first adventurous seed then floated for weeks on the swift Humboldt Current before washing ashore on a rocky Galapagos beach. From there it may have proceeded inland the way Darwin envisioned, wafted by an onshore breeze. But there is another possibility that is just as likely and far more charming. Throughout the arid lowlands of the Galapagos, endemic finches line their nests exclusively with seed plumes, making it possible that Darwin's cotton made the last stage of its journey in the beak of a Darwin's finch.

The odds of any particular seed finding a good home by wind or wave seem long. But given time and repetition, both strategies produce results. More plants rely on wind for dispersal than all other methods combined, although usually at a scale of inches or feet. Add an ocean current to the equation, however, and stories like Darwin's cotton become practically commonplace—at least 170 other plant species arrived in the archipelago by similar means. In fact, reaching the Galapagos hardly sounds like a feat at all considering how cotton colonized South America in the first place—crossing the full breadth of the Atlantic Ocean not once, but twice. Biogeographers call it "a miracle squared," but the evidence is unequivocal. American cotton species contain the genes of two distinct African ancestors, adding an evolutionary twist to a transatlantic relationship with impacts far beyond the interest of botanists. In the nineteenth century, the movement of cotton across the Atlantic Ocean lay at the heart of world events—industrialization, globalization, British ascendancy, slavery, and the American Civil War.

It's hard to overstate the importance of cotton in shaping the modern era. Historians have called it "the revolutionary fiber," and "the fuel of the industrial revolution." It became the first global, mass-produced commodity and anchored an infamous "trade triangle" that connected American plantations, British mills, and African slave ports—raw cotton flowed east, finished fabrics flowed south, and slave labor flowed west. As Karl Marx put it, "without slavery you have no cotton, without cotton you have no modern industry." Marx wrote those words in 1846, a time when the cotton trade accounted for a staggering 60 percent of American exports and employed one out of five British workers. In raw and finished forms, it remained the dominant European and American export activity for more than a century. But the sweeping social and economic changes wrought by cotton seed fibers began far earlier.

When Christopher Columbus encountered cotton in the Caribbean, he naturally took it as further evidence that he'd reached the coast of Asia. For over a thousand years, cotton had been considered a distinctly Asian fabric, produced in India and distributed along trade routes that stretched east to Japan and west as far as Africa and the Mediterranean. Persia alone imported between 25,000 and 30,000 camel loads of Indian cotton every year, and a modest but steady supply reached Europe by way of Venice, where it was seen as a lucrative supplement to the spice trade. Cotton was bought and sold widely within Asia as well. Historians have often noted that the Silk Road was a Cotton Road when viewed in reverse. Chinese merchants returned home with vast amounts of the Indian fabric, but still couldn't keep up with demand. Eventually, China created its own supply by decree—a strict fourteenth-century law required anyone farming more than an acre to plant part of that land in cotton. When Portuguese and Dutch ships first reached Asian ports in search of spices, they found cotton to be a vital part of the equation. Printed fabrics from India often had more purchasing power than European silver, especially when trading with the remote islands where nutmeg and cloves were grown. Textiles went on to become a profitable sideline for the Dutch, but it was the British East India Company that really ushered in the new cotton era.

During the latter half of the eighteenth century, three things aligned to transform the economics of cotton: fashion, innovation, and politics. By copying the designs of expensive silks at a fraction of the cost, calicoes (from the coastal city of Calicut) and other printed fabrics helped introduce color and a sense of style to Europe's growing middle class. In spite of resistance from the wool and linen industries—including protectionist laws, the occasional fabric riot, and the surprising spectacle of calico-clad women being attacked and stripped naked in the streets of London—imports of Indian cotton boomed. The East India Company shifted its trade from spices to textiles, feeding markets not only in Europe but also in British-held territories around the world, from Africa to Australia to the West Indies. The success of Indian fabrics as a global commodity inspired imitation and led to a series of transformative inventions: James Hargreaves's spinning jenny, Samuel Crompton's spinning mule, and Richard Arkwright's water frame. Mechanization increased the quality of British-made fabrics and dropped the price, shifting global production from Indian villages to English mill towns. The Industrial Revolution was underway, with the seeds of cotton inspiring its machinery just as another seed, coffee, had revved up the minds of its workforce.

Politically, rising demand for cotton—and the need for a steady supply—helped justify British expansion in India. Through coercion and conquest, the East India Company came to dominate the subcontinent at the same time that British mills were stealing away the cotton business, undermining India's economy. It's no wonder Mahatma Gandhi chose homespun cotton as a symbol of resistance to British rule, saying it was "the patriotic duty of every Indian to spin his own cotton and weave his own cloth." A stylized spinning wheel remains the focal point of the Indian national flag. As the first highly mechanized industry, cotton helped shift Europe from an economy of farms to one of factories, establishing a pattern that would hold for two centuries—the import of raw materials from south to north, followed by the export of finished products to the world. In Europe, that system bolstered empires and created a surge of prosperity. In America, it led to war.

The cotton encountered by Christopher Columbus in the New World differed from its African and Asian relations. The fibers were longer and the seeds stickier, making it notoriously difficult to work with. But the good admiral praised his cotton nonetheless, claiming that it grew in abundance, required no tending, and could be harvested year round. His descriptions were full of typical Columbus braggadocio, but in the case of cotton his enthusiasm wasn't far from the mark. Longer fibers in the plumes made for a superior yarn, and a single species of American cotton now accounts for over 95 percent of world production. But as my own attempts made clear, separating seeds from fiber was no easy task. In spite of the worldwide boom, American cotton remained a minor crop until Eli Whitney assembled his famous ginning machine. It sparked immediate increases in efficiency and productivity, but there is no way the young inventor could have predicted the other consequences that lay in store.

Though he received a patent for the device signed by then secretary of state Thomas Jefferson (who ordered a gin for Monticello after looking at the plans), Eli Whitney never profited from his invention. Its simple design made it simple to copy, and he soon learned that rural southern courts had little compassion for an urban northern patent holder. Collecting even a small portion of his due would have brought Whitney phenomenal wealth. In the decade following his fruitless patent, cotton-gin technology drove a fifteen-fold increase in exports from the American South. Production continued to double every decade, and by the mid-nineteenth century, southern plantations accounted for nearly three-quarters of the world's raw cotton supply. More than any other commodity, it gave the young American nation wealth, clout, and international prestige.

No historian disputes the legal woes of Eli Whitney, but those wrongs pale in comparison to the other consequences of his invention. Mechanization may have simplified the processing of cotton, but growing the crop still demanded massive inputs of labor. The suddenly profitable American cotton business reinvigorated what had been a declining market for African slaves. That most gruesome third leg of the Atlantic trade triangle surged to a new peak in the 1790s, when as many as 87,000 slaves crossed the Middle Passage to America every year. The US Congress banned human trafficking from abroad in 1808, but the domestic trade continued to flourish, and the number of slaves quintupled between 1800 and 1860. In some places, buying and selling the people who picked cotton became a business that rivaled the buying and selling of cotton itself.

This deep-rooted coupling of slavery and cotton came to define the economy of the Antebellum South, setting the stage for America's deadliest conflict. Over a million people would be killed, wounded, or displaced by the time the Civil War ended in 1865. It was a defining confrontation that left behind persistent social and political divisions. But the underlying economics of seed fluff hardly changed at all. With sharecropping in place of slavery, cotton production rebounded to prewar levels within five years, and it remained the top American export until 1937. Things worked out fine for Eli Whitney, too. His cotton-gin patent expired, still worthless, but he went on to make a fortune in a different industry—the manufacture of muskets, rifles, and pistols. Ironically, weapons from the Whitney Armory were among the most common firearms used in the Civil War.

While seeds and warfare may sound like odd bedfellows, cotton fluff is not the only dispersal strategy to influence events on a battlefield. The first aerial bombardment in history consisted of four small hand grenades chucked from the cockpit of a reconnaissance plane during the Italo-Turkish War of 1911. The Italian pilot acted alone, without orders, and pulled the pins himself as he swooped low over a Turkish encampment in the Libyan Desert. There were no injuries, but people on both sides of the conflict decried his action as a shocking breach of military etiquette. That sense of outrage passed quickly, however, as strategists recognized the potential of this new kind of attack. Dropping those grenades ushered in a new era in war-making, and the pilot earned himself a permanent footnote in textbooks of military history. But few people remember the unusual design of his aircraft. It was not a biplane in the style of the Wright brothers or Alberto Santos-Dumont, the Brazilian pioneer in aviation; nor was it inspired by the bird-like gliders of Otto Lilienthal. It consisted of a tail fan and a shapely wing that would have looked familiar to anyone from Indonesia, where the same design wafted about by the thousands in the canopy of the rainforest. The airplane used in that famous incident was essentially a flying seed, scaled-up from the streamlined pips of a Javan cucumber.

Most pioneering aviators took their cues from birds and bats, but Austrian Igo Etrich looked to a far older version of wings. Fossils like those I saw with Bill DiMichele show that seeds have been winged for hundreds of millions of years. With that amount of practice, plants have thinned and stretched the tissues of their offspring into a huge diversity of fins and props, from the honeycombed ridges found on "fringed grass of Parnassus" to the layered skirts of a larkspur, or to the more familiar whirligigs of maple and ash. Etrich focused his attention on a seed whose single backswept wing reached six inches (fifteen centimeters) across, but, like a cotton fiber, was only one cell thick, providing lift without weight. The vine of the Javan cucumber is thin and nondescript, tangling upward toward the sunlit treetops of Indonesian forests. Few Western botanists have seen one—but they knew about the seeds long before they found the plant that made them.

Drifting down by the hundreds from the split end of every pumpkin-shaped fruit, Javan cucumber seeds commonly fly long distances from the edges of the rainforest. Sailors have reported finding

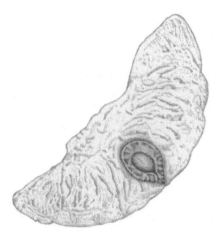

FIGURE 13.3. Javan cucumber (*Alsomitra macrocarpa*). With its edges stretched into a broad, thin wing, the seed of the Javan cucumber is one of nature's most efficient airfoils, staying aloft on the slightest breeze and gliding for trips measured not in feet, but in miles. ILLUSTRATION © 2014 BY SUZANNE OLIVE.

them on the decks of ships, miles out to sea. The seeds accomplish these extraordinary glides through traits attractive to any aeronautical engineer—passive stability and a shallow angle of descent. The first term refers to self-correction in flight, the ability to regain equilibrium when things start to wobble. The flexible membrane of a Javan cucumber seed achieves this stability inherently, with contours that constantly readjust the center-point of lift. Having a shallow descent angle means that the seeds lose less than a foot and a half (half a meter) in altitude for every second they fly. (Twirling maple seeds, by comparison, fall at more than twice that rate.) Though he called his plane the *Taube*, German for "Dove," Etrich didn't hide the fact that it was based on a seed, and Javan cucumbers have enjoyed a cult following in aviation circles ever since. Production *Taubes* featured a tailed fuselage dividing the curve of the wing, but

FIGURE 13.4. The Northrop Grumman B-2 Spirit, better known as the Stealth Bomber, took inspiration from the flying-wing design of Javan cucumber seeds. WIKIMEDIA COMMONS.

Etrich dreamed of building planes that mimicked the cucumber seed exactly—eliminating the tail and fitting the cockpit inside a single, uninterrupted wing surface. Mainstream aviation moved away from that vision after World War I, but the idea of a "flying wing" aircraft persisted in the imaginations of a few maverick designers for the next seventy-five years, culminating in what is still considered the most advanced, expensive, and deadly airplane ever built.

The Northrop Grumman B-2 Spirit is better known by its popular name, the Stealth Bomber. Like the seed that inspired it, the B-2's high-lift, low-drag shape is extremely efficient, making it possible for the plane to fly nearly 7,000 miles (11,265 kilometers) without refueling. As an added bonus, its lack of a tail or other protruding fins helps the B-2 avoid detection by all but the most advanced air defense systems. Though only twenty-one of them have ever been built (at a total program cost exceeding $2 billion per plane), the B-2 is considered a cornerstone of the US arsenal. Designed for both nuclear and conventional payloads, a single plane has the capacity to extinguish more lives than the entire Civil War. It's an impressive feat of engineering, but seems ill-matched for the Javan cucumber's flying wing, a design that evolved not to end life, but to spread it.

Military aviation is a notoriously competitive industry, and any number of defense contractors would love to replace Northrop Grumman's plane with their own. The Stealth Bomber's seed-inspired wing gives it an edge, just enough aerodynamic advantage to out-perform its rivals and keep the B-2 program funded. That kind of constant competition helps aircraft design continue to evolve, but it also begs an obvious question about seed dispersal—Which is better, wing or plume? With an eight-foot stepladder, a measuring tape, and an enthusiastic, seed-loving preschooler, it seemed like a question I was well equipped to answer.

On a warm summer morning, Noah and I headed down to our back field to play a game we called "seed drop." The rules were simple. I climbed the ladder and let fly with a series of seeds, while Noah chased them down across the field, planting orange survey flags wherever one drifted to earth. Then we took the measuring tape and saw how far they'd traveled. We started with cotton, but it proved to be something of a disappointment. Even in a moderate breeze, the fluff wafted no more than 16 feet (5 meters) before falling into the grass. That's better than going straight down, but it would obviously take quite a gale to get cotton seeds across an ocean. Their fibers may be light and prolific, but the seeds themselves remain rather bulky, leading some experts to think flotation may be their most important evolutionary advantage. (It's telling that, like Darwin's cotton, the fluffiest varieties all occur near coastlines.) We moved on to dandelions (30 feet [9 meters]), and then tried the plumes from a nearby poplar tree, which drifted all the way to the edge of the forest, 115 feet (35 meters) away.

When it came time to test something winged, I carefully pulled a single Javan cucumber seed from an envelope. (Although my letters to a botanical garden near Jakarta had gone unanswered, I'd managed to purchase a few from a sympathetic seed collector.) It was easy to see why Igo Etrich had been inspired. As large as my outstretched hand, and thin as a leaf, the seed consisted of a thumb-sized golden disc surrounded by a translucent membrane that crinkled in the breeze like parchment. It looked unbeatable and eager to soar, but the first flight was a total failure.

"That's no good at all, Papa," Noah said with obvious disgust, as the seed wobbled and crashed after a few feet, like a paper airplane with a nose-heavy fold. Five more times I climbed to the top rung and let it go, but only once did the seed really start to fly, dipping and swerving like a nervous bird for nearly 50 feet (15 meters). That was farther than cotton, but hardly seemed like a good model for a billion-dollar airplane. I could see Noah's interest beginning to wane as I climbed the ladder one last time. Again, the seed dipped and fluttered a bit as it descended toward the grass. Then, as if tugged by invisible intent, it caught just the right breeze and wafted suddenly upward. Noah whooped, I raced down the ladder, and the chase was on.

We ran to the edge of the field, following the seed as it soared, turned, and cleared the orchard fence in a series of long swoops and lurches. One of the barn swallows nesting nearby in the eaves of the Raccoon Shack flew over to investigate, circling the seed twice as it continued its ascent. Laughing in sheer wonderment, we watched that seed clear the top of the forest, where it joined a current of faster air and started picking up speed.

"There it goes, Noah!" I heard myself shouting. "We'll never see it again!"

He still had his orange flag and I was holding the measuring tape, but our contest between wings and plumes was forgotten. We watched that seed fly for the simple joy of seeing something beautiful doing what it is meant to do. Standing there together, heads tilted skyward, we laughed and laughed until it disappeared from view—a papery wisp at the edge of vision, still rising.

The Future of Seeds

By cosmic rule,
as day yields night,
so winter summer,
war peace, plenty famine.
All things change.

—Heraclitus (sixth century BC)

E very family has its lore. My father's ancestors all hailed from
Norway—stoic fisher-folk who plied the fjords in small wooden
boats. Nowadays, none of us makes a living with hook and line, but
fishing remains an expected activity. It's hard to find a family photo
where someone isn't holding a dead salmon. My mother called her
side "Heinz 57," a typical American mish-mash that included everything from country doctors and a horse thief to a congressman killed
in a duel. When I married Eliza, I joined a clan with a strong farming tradition. I'm still learning her family's stories, but many of them
seem to involve watermelons.

"I grew the first tetraploid watermelon in North America," Eliza's
grandfather, Robert Weaver, told me with a merry gleam in his eye.
At ninety-four, Bob hasn't lost one whit of his verve for life, and still
recalls every detail of the melon years—hand-pollinating countless

blossoms, and trying in vain to sell the results. "I went to see the Burpee boys," he said, referring to the brothers who ran the famous seed company, "but they didn't know a chromosome from dirt!"

The word *tetraploid* refers to the number of chromosomes in the nucleus of a cell. As Gregor Mendel learned, plants typically have two sets of chromosomes, one from each parent, a condition geneticists call *diploid*. But sometimes cell division goes wrong, producing seeds with double the normal number. In nature, this is an important source of variation and can lead to new traits, varieties, and species. *Almendro* is a tetraploid, and so is Darwin's cotton. But in the mid-twentieth century, plant breeders discovered that chromosomes could be doubled chemically, and that back-crossing a tetraploid with the diploid parent produced infertile hybrids.* The upshot for watermelons was a normal-looking, sweet-tasting fruit incapable of filling itself with seeds. For consumers, it offered convenience, and for seed companies it offered control, since farmers and gardeners would have to buy new seeds every year rather than saving their own.

Today, seedless varieties account for over 85 percent of the watermelon market, but Bob sold his shares of the family enterprise decades before it showed a profit. "Millions," he told me, when I asked how much his brother-in-law had eventually made from the idea. But he said it without a hint of regret. Bob walked away from the melon business to move his family west, settling on an island where, as he put it, "the kids could walk barefoot to school." They built a house from driftwood logs overlooking a garden with soil so rich he once harvested twenty-four pounds of potatoes from a single plant.

---

*The crux of the matter is simple division. Plants with an even number of chromosomes can easily give half that allotment to their pollen or sperm cells, which then unite to form a seed. Crossing diploid and tetraploid plants, however, produces individuals with three sets of chromosomes, a number that cannot be divided evenly. Triploid plants may be healthy, but they are sterile, unable to make viable pollen or eggs, and therefore unable to form seeds.

In many ways, Bob's experience foreshadowed the controversy now raging about the future of seeds. He grew up farming, and later returned to a simple rural lifestyle. But in between he glimpsed the beginnings of genetic modification, a field that now goes far beyond the simple doubling of chromosomes. Modern plant geneticists have the tools to add, delete, alter, transfer, and potentially create specific genes for specific traits. The possibilities are endless, but also troubling. Farmers now face patent disputes over seed saving, open pollination, and other age-old traditions, and critics have raised legitimate worries about the environmental, health, and even moral implications of mixing up genes from different species. Genetically modified seeds have joined a growing list of technologies and innovations that we struggle to make peace with, from drones to cloning to nuclear power. Some people embrace the idea of manipulating seeds, particularly those who stand to profit, but many others are wary or have turned away altogether. There can be no single solution that satisfies everyone, but if you've read this far, then you, too, have thought a great deal about seeds, and I hope we can all agree on one point: they are worthy of the debate.

Bob Weaver never farmed commercially again, but gardening remained a central part of his family's life. He and his wife passed the love of it on to their kids, who passed it on to Eliza's generation. And it now has a firm grip on Noah. Whenever the Weaver family gathers, conversation eventually turns to who is growing what, and how it's doing. It's not unusual for seed packets to emerge, prompting scribbled notes, the folding of ad hoc envelopes, and an immediate exchange of promising varieties. Like the Mende people of Sierra Leone "trying out new rice," gardeners everywhere share an eagerness to trade seeds, constantly experimenting in whatever plot of earth they till. And through that tradition of gift and receipt, the seeds take on stories.

When I interviewed Diane Ott Whealy at the Seed Savers Exchange, she told me that planting her grandfather's morning glory was like having him with her. All summer long she saw him there, in the purple blossoms winking up from a hedgerow, or peering out of the greenhouse. So it is in our garden, where in any given year Eliza might plant the Storage #4 cabbage her grandfather swears by, or her Aunt Chris's kale (which came to Chris from a one-legged Scots-Irishman named McNaught). Or perhaps she will sow the family's favorite pole bean, Oregon Giant—hard to come by now unless you save the seed yourself. In exchange, I like to think that one of her relations is growing "Eliza's lettuce," a local strain of the Salad Bowl variety that she discovered years ago, gone to seed near her plot in a community garden.

Evolution behaves much like a gardener, saving only the most successful experiments. And there is nothing necessarily inviolate about the triumph of seeds. Just as the spore plants yielded their dominance, so might seeds eventually give way to something new. In fact, this process may already be underway. With over 26,000 recognized species, orchids make up the most diverse and highly evolved plant family on earth. Yet their seeds are hardly seeds at all. Break open an orchid pod and the seeds puff out like dust, microscopic blips that essentially lack seed coats, defensive chemicals, or any discernable nutrition. They are still baby plants, but in Carol Baskin's analogy, they don't have the box or the lunch. In fact, they can only germinate and grow if they land in soil that contains just the right kind of symbiotic fungi. As such, orchid seeds have very little to offer people—no fuels, fruits, foods, or fibers, no stimulants or useful drugs. Only one species out of all those thousands produces a seed product with any commercial value—vanilla. If orchids didn't have beautiful flowers, we would hardly be aware of them at all.

Paleobotanists like Bill DiMichele take the long view on plant evolution, watching traits, species, and whole groups wax and wane through the fossil record. Bill doesn't think the reign of seeds will end anytime soon. "Orchids are freeloaders," he assured me. In addition to their reliance on fungi, most species are epiphytes, using other plants for support and structure. And their attractive flowers rarely contain nectar or accessible pollen, a system of trickery that would fall apart if other plants didn't offer reliable rewards. Still, if nearly one out of every ten species in the global flora is an orchid, it's hard not to believe that they're on to something. The success of using simple, dust-like seeds reminds us that complexity is a symptom of evolution, not a consequence. All the elaborate and remarkable features found in seeds—from nourishment to endurance to protection—will persist only so long as they benefit future generations. Seeds embody the biology of passing things down. In a sense, that is also the root of their deep cultural significance. Seeds give us a tangible connection from past to future, a reminder of human relationships as well as the natural rhythms of season and soil.

Last fall, Noah and I collected the seeds of bellflower and pink mallow from my mother's overgrown flower garden. I brought them home to pep up a patch of bare ground in front of the Raccoon Shack. On an afternoon in early spring, we shoveled up the soil of our small plot, pulled a few weeds, and brought out the seeds. Noah inspected them closely, commenting on the blackish nubs of mallow in their webbed pouches, and the tiny bellflowers, like flecks of golden dust. When it came time to plant, he threw exuberant fistfuls across the turned earth, and then added something of his own—four kernels of popcorn carefully saved from a snack earlier in the day.

As luck would have it, we chose a perfect moment for planting. Steady rain that afternoon watered in the seeds and then the skies cleared, giving us days and days of sunshine. The mallows germinated quickly, surging upward with wisps of seed coat still clinging to their new leaves. Two weeks have passed, and as I write these lines I can hear Eliza outside my office window, pointing out the baby plants to Noah. "There's another *malva*," she tells him, using the botanical name. "See it?"

He answers yes, and later shows the seedlings proudly to me— brave green specks brightening a backdrop of dirt. By the time this book goes to print, they will be in bloom.

Appendix A
Common and Scientific Names

The following list includes common, scientific, and family names for all plant species mentioned in the text.

COMMON NAME	SCIENTIFIC NAME	FAMILY
Acacia	*Acacia* spp.	Fabaceae (bean)
Adzuki bean	*Vigna angularis*	Fabaceae (Bean)
Afzelia	*Afzelia africana*	Fabaceae (Bean)
Almendro	*Dipteryx panamensis*	Fabaceae (Bean)
Almond	*Prunus dulcis*	Rosaceae (Rose)
Apple	*Malus domestica*	Rosaceae (Rose)
Arm millet	*Brachiaria* spp.	Poaceae (Grass)
Asparagus	*Asparagus officinalis*	Asparagaceae (Asparagus)
Aster	*Aster* spp.	Asteraceae (Aster)
Autumn crocus	*Colchicum autumnale*	Colchicaceae (Autumn Crocus)
Avocado	*Persea americana*	Lauraceae (Laurel)
Balanites	*Balanites wilsoniana*	Zygophyllaceae (Caltrop)
Barley	*Hordeum vulgare*	Poaceae (Grass)
Basil	*Ocimum basillicum*	Lamiaceae (Mint)

continues

COMMON NAME	SCIENTIFIC NAME	FAMILY
Bishop's flower	*Ammi majus*	Apiaceae (Parsley)
Bent-grass	*Agrostis* spp.	Poaceae (Grass)
Bitterbark	*Sacoglottis gabonensis*	Humiriaceae (Humiria)
Blackbean	*Castanospermum australe*	Fabaceae (Bean)
Blackberry	*Rubus* spp.	Rosaceae (Rose)
Black currant	*Ribes nigrum*	Grossulariaceae (Gooseberry)
Blueberry	*Vaccinium* spp.	Ericaceae (Heather)
Bluegrass	*Poa* spp.	Poaceae (Grass)
Bodhi fig	*Ficus religiosa*	Moraceae (Mulberry)
Burdock	*Arctium* spp.	Asteraceae (Aster)
Cacao	*Theobroma cacao*	Malvaceae (Mallow)
Calabar bean	*Physostigma venenosum*	Fabaceae (Bean)
Canadian fleabane	*Conyza canadensis*	Asteraceae (Aster)
Canary grass	*Phalaris* spp.	Poaceae (Grass)
Canna lily	*Canna indica*	Cannaceae (Canna Lily)
Canola (rape)	*Brassica napus*	Brassicaceae (Mustard)
Carob	*Ceratonia siliqua*	Fabaceae (Bean)
Cashew	*Anacardium occidentale*	Anacardiaceae (Cashew)
Cassia	*Cinnamomum cassia*	Lauraceae (Laurel)
Castor bean	*Ricinus communis*	Euphorbiaceae (Spurge)
Celery	*Apium graveolens*	Apiaceae (Parsley)
Cheat grass	*Bromus tectorum*	Poaceae (Grass)
Chestnut	*Castanea* spp.	Fagaceae (Beech)
Chickpea (garbanzo bean)	*Cicer arietinum*	Fabaceae (Bean)
Chigua	*Zamia restrepoi*	Zamiaceae (Cycad)
Chili pepper	*Capsicum* spp.	Solanaceae (Nightshade)
Chinese sicklepod	*Senna obtusifolia*	Fabaceae (Bean)
Climbing oleander	*Strophanthus gratus*	Apocynaceae (Dogbane)

continues

COMMON NAME	SCIENTIFIC NAME	FAMILY
Coconut	*Cocos nucifera*	Arecaceae (Palm)
Coffee	*Coffea* spp.	Rubiaceae (Madder)
Congo jute	*Urena lobata*	Malvaceae (Mallow)
Coral bean	*Adenanthera pavonina*	Fabaceae (Bean)
Corn (maize)	*Zea mays*	Poaceae (Grass)
Cotton	*Gossypium* spp.	Malvaceae (Mallow)
Cowpea	*Vigna unguiculata*	Fabaceae (Bean)
Cranberry	*Vaccinium* spp.	Ericaceae (Heather)
Cucumber	*Cucumis sativus*	Cucurbitaceae (Gourd)
Cycad	*Cycas* spp.	Cycadaceae (Cycad)
Dandelion	*Taraxacum officinale*	Asteraceae (Aster)
Darwin's cotton	*Gossypium darwinii*	Malvaceae (Mallow)
Date palm	*Phoenix dactylifera*	Arecaceae (Palm)
Dwarf mallow	*Malva neglecta*	Malvaceae (Mallow)
Eggplant	*Solanum melongena*	Solanaceae (Nightshade)
False hellebore	*Veratrum viride*	Melanthiaceae (Bunchflower)
Feather grass	*Stipa* spp.	Poaceae (Grass)
Fescue	*Festuca* spp.	Poaceae (Grass)
Fig	*Ficus* spp.	Moraceae (Mulberry)
Forget-me-not	*Myosotis* spp.	Boraginaceae (Borage)
Frankincense	*Boswellia sacra*	Burseraceae (Torchwood)
Fringed grass of Parnassus	*Parnassia fimbriata*	Celastraceae (Staff Tree)
Garbanzo bean (chickpea)	*Cicer arietinum*	Fabaceae (Bean)
Ginkgo	*Ginkgo biloba*	Ginkgoaceae (Ginkgo)
Goat grass	*Aegilops* spp.	Poaceae (Grass)
Gorse	*Ulex* spp.	Fabaceae (Bean)
Groundnut	*Vigna subterranean*	Fabaceae (Bean)
Guar	*Cyamopsis tetragonoloba*	Fabaceae (Bean)

continues

COMMON NAME	SCIENTIFIC NAME	FAMILY
Hairy panic grass	*Panicum effusum*	Poaceae (Grass)
Hawkweed	*Hieracium* spp.	Asteraceae (Aster)
Hawthorn	*Cratageous* spp.	Rosaceae (Rose)
Hazel	*Corylus* spp.	Betulaceae (Birch)
Henbane	*Hyoscyamus niger*	Solanaceae (Nightshade)
Hibiscus	*Hibiscus* spp.	Malvaceae (Mallow)
Holly	*Ilex* spp.	Aquifoliaceae (Holly)
Hollyhock	*Alcea* spp.	Malvaceae (Mallow)
Horse chestnut	*Aesculus hippocastanum*	Sapindaceae (Soapberry)
Horse-eye bean	*Ormosia* spp.	Fabaceae (Bean)
Indian lotus	*Nelumbo nucifera*	Nelumbonaceae (Lotus)
Iris	*Iris* spp.	Iridaceae (Iris)
Javan cucumber	*Alsomitra macrocarpa*	Cucurbitaceae (Gourd)
Jojoba	*Simmondsia chinensis*	Simmondsiaceae (Jojoba)
Junglesop	*Anonidium mannii*	Annonaceae (Custard Apple)
Kale	*Brassica oleracea*	Brassicaceae (Mustard)
Kola nut	*Cola* spp.	Malvaceae (Mallow)
Larkspur	*Delphinium* spp.	Ranunculaceae (Buttercup)
Lentil	*Lens culinaris*	Fabaceae (Bean)
Madrona	*Arbutus menziesii*	Ericaceae (Heather)
Maize (corn)	*Zea mays*	Poaceae (Grass)
Manzanillo	*Hippomane mancinella*	Euphorbiaceae (Spurge)
Maple	*Acer* spp.	Sapindaceae (Soapberry)
Marula	*Sclerocarya birrea*	Anacardeaceae (Cashew)
Maté	*Ilex paraguariensis*	Aquifoliaceae (Holly)
Maygrass	*Phalaris caroliniana*	Poaceae (Grass)
Milk thistle	*Silybum marianum*	Asteraceae (Aster)
Mistletoe	*Viscum* spp.	Viscaceae (Mistletoe)
Moth mullein	*Verbascum blatteria*	Scrophulariaceae (Figwort)
Mulga grass	*Aristida contorta*	Poaceae (Grass)

continues

COMMON NAME	SCIENTIFIC NAME	FAMILY
Mung bean	*Vigna radiata*	Fabaceae (Bean)
Naked woollybutt	*Eragrostis eriopoda*	Poaceae (Grass)
Nardoo	*Marsilea* spp.	Marsileaceae (Water-clover)
Nutmeg	*Myristica fragrans*	Myristicaceae (Nutmeg)
Oak	*Quercus* spp.	Fagaceae (Beech)
Oil palm	*Elaesis guineensis*	Arecaceae (Palm)
Okra	*Abelmoschus esculentus*	Malvaceae (Mallow)
Omwifa	*Myrianthus holstii*	Urticaceae (Nettle)
Paintbrush	*Castilleja* spp.	Orobanchaceae (Broomrape)
Pea	*Pisum sativum*	Fabaceae (Bean)
Pepper (black or white)	*Piper nigrum*	Piperaceae (Pepper)
Pepper (chili)	*Capsicum* spp.	Solanaceae (Nightshade)
Pincushion protea	*Leucospermum* spp.	Proteaceae (Protea)
Poison hemlock	*Conium maculatum*	Apiaceae (Parsley)
Poplar	*Populus* spp.	Salicaceae (Willow)
Quandong	*Santalum acuminatum*	Santalaceae (Sandalwood)
Quince	*Cydonia oblonga*	Rosaceae (Rose)
Rape (canola)	*Brassica napus*	Brassicaceae (Mustard)
Raspberry	*Rubus* spp.	Rosaceae (Rose)
Ray grass	*Sporobolus actinocladus*	Poaceae (Grass)
Rock rose	*Cistus* spp.	Cistaceae (Rock Rose)
Rosary pea	*Abrus precatorius*	Fabaceae (Bean)
Silk tree	*Albizia julibrissin*	Fabaceae (Bean)
Soapwort	*Saponaria officinalis*	Caryophyllaceae (Carnation)
Sorghum	*Sorghum* spp.	Poaceae (Grass)
Soybean	*Glycine max*	Fabaceae (Bean)
Squash	*Cucurbita* spp.	Cucurbitaceae (Gourd)
Star grass	*Dactyloctenium radulans*	Poaceae (Grass)
Sugarcane	*Saccharum* spp.	Poaceae (Grass)

continues

COMMON NAME	SCIENTIFIC NAME	FAMILY
Suicide tree	*Cerbera odollam*	Apocynaceae (Dogbane)
Sumac	*Rhus* spp.	Anacardiaceae (Cashew)
Sweet clover	*Melilotus* spp.	Fabaceae (Bean)
Sycamore	*Acer pseudoplatanus*	Sapindaceae (Soapberry)
Tagua	*Phytelaphas* spp.	Arecaceae (Palm)
Tara	*Caesalpinia spinosa*	Fabaceae (Bean)
Tea	*Camellia sinensis*	Theaceae (Tea)
Tomato	*Solanum* spp.	Solanaceae (Nightshade)
Tonka bean	*Dipteryx odorata*	Fabaceae (Bean)
Tsamma melon (watermelon)	*Citrullus lanatus*	Cucurbitaceae (Gourd)
Velvet bean	*Mucuna pruriens*	Fabaceae (Bean)
Vernal grass	*Anthoxanthum odoratum*	Poaceae (Grass)
Vetch	*Vicia* spp.	Fabaceae (Bean)
Wallace's spike moss	*Selaginella wallacei*	Sellaginelaceae (Spike Moss)
Watermelon (*tsamma* melon)	*Citrullus lanatus*	Cucurbitaceae (Gourd)
Wheat	*Triticum* spp.	Poaceae (Grass)
Wild oat	*Avena* spp.	Poaceae (Grass)
Willow	*Salix* spp.	Salicaceae (Willow)
White hellebore	*Veratrum album*	Melanthiaceae (Bunchflower)
Yew	*Taxus* spp.	Taxaceae (Yew)

Appendix B
Seed Conservation

A portion of the proceeds from this book will be donated to help preserve the diversity of seeds from both wild and cultivated species. To further these efforts directly, consider making a donation to one of the following organizations.

Seed Savers Exchange
   3094 North Winn Road
   Decorah, IA 52101, USA
   Phone: (563) 382–5990
   www.seedsavers.org

Organic Seed Alliance
   PO Box 772
   Port Townsend, WA 98368, USA
   Phone: (360) 385–7192
   www.seedalliance.org

Global Crop Diversity Trust
   Platz Der Vereinten Nationen 7
   53113 Bonn, Germany
   Phone: +49 (0) 228 85427 122
   www.croptrust.org

The Millennium Seed Bank Partnership
Royal Botanic Gardens, Kew
Richmond, Surrey TW9 3AB, UK
Phone: +44 020 8332 5000
www.kew.org

MILLENNIUM
SEED BANK
PARTNERSHIP
Kew

Notes

Preface: "Heed!"

xvii **breadfruit seedlings overboard:** While the HMS *Bounty* gained
fame for its mutiny, the purpose of the journey was botanical. At
the suggestion of Sir Joseph Banks, president of the Royal Society,
Captain Bligh's orders involved the transport of live breadfruit
trees from their native Tahiti to the West Indies, where plan-
tation owners hoped they would yield cheap sustenance for the
growing slave population. After his eventual return to England,
Bligh set forth again in the HMS *Providence* and completed his
original commission, delivering over 2,000 healthy saplings to Jamaica. Although the trees thrived in their new home, the scheme
failed in light of one detail that had been overlooked: African
slaves found Polynesian breadfruit disgusting and refused to eat it.

xviii **addressed only briefly in these pages:** For insightful analyses on
genetically modified crops, see Charles 2002, Cummings 2008,
and Hart 2002.

Introduction: The Fierce Energy

xxiv **"just as the little boy had known it would":** Krauss 1945.

xxiv **353,000 other kinds of plants that use seeds to reproduce:** Estimates of the number of seed plants vary from 200,000 to over
420,000 (Scotland and Wortley 2003). The figure used here
comes from an ongoing collaboration among the world's largest herbaria, including Kew Gardens, the New York Botanical

Garden, and the Missouri Botanical Garden (The Plant List 2013, Version 1.1, archived at www.theplantlist.org).

CHAPTER ONE: SEED FOR A DAY

3 **length of its own body:** Herpetologists with high-speed cameras have repeatedly shown that viper strikes reach only about a third or a half of their body length (e.g., Kardong and Bels 1998). Even well-informed observers, however, often give wildly exaggerated accounts of snake strikes (see Klauber 1956 for great examples). Having seen a fer-de-lance (*Bothrops asper*) in action, I'll throw my lot in with the hyperbolists—anything seems possible when those fangs are headed in your direction.

6 **rainforest in the first place:** A member of the pea family, *almendro* is known to science as *Dipteryx panamensis* (also *D. oleifera*). With apologies for the self-reference, see Hanson et al. 2006, 2007, and 2008 for more on the *almendro*'s role as a keystone species in Central American rainforests.

7 **rapidly developing rural landscape:** I also had an ulterior motive. Before my *almendro* study, I had worked on projects involving mountain gorillas and brown bears, species known in the trade as "charismatic megafauna," the big fancy animals. In the *almendro*, the plant lover inside me saw a chance to promote "charismatic megaflora." What better way to describe a 150-foot keystone tree with iron-hard wood and Marge Simpson's haircut?

9 **rainforests of southern Mexico and Guatemala:** The avocado tree (*Persea americana*) is known only as a cultivated species. Sometime in the thousands of years since domestication, its wild ancestor disappeared from the forests of Central America. One theory suggests that many large-fruited neo-tropical trees faded away following the loss of their seed dispersers: giant armadillos, glyptodonts, mammoths, gomphotheres, and other extinct Pleistocene megafauna (Janzen and Martin 1982). With its massive seed, the wild avocado would certainly have required the services of a large-bodied animal to move it around. (Of course, people now play that role quite well, and avocados can be found growing on every continent except Antarctica!)

11 **always right for sprouting:** Botanists call seeds that don't survive desiccation *recalcitrant*. Though rare in temperate and seasonal climates, this strategy is found in an estimated 70 percent

of tropical rainforest trees, where quick germination offers more of an advantage than long-term dormancy. What works in a jungle, however, makes things difficult in a storage facility. Christina Walters at the US National Seed Bank calls recalcitrant seeds "spoiled little children," but has found some success flash-freezing isolated embryos in liquid nitrogen.

12   **tenth of an ounce I've ever measured:** Because they never dry out and enter a truly desiccated, dormant state, avocado seeds take up only a small amount of water, essentially participating only in the last stage of full imbibition. Dry seeds generally imbibe two to three times their weight in water.

14   **stifling effects of the caffeinated bean:** Cell division in plants takes place in specialized tissue called *meristem*, which is located primarily at the tips of growing roots and stems. Once the coffee's expanding cells push these tips out and away from the caffeine, the meristem can divide, and growth through division begins—a tidy system.

15   **" . . . season at which each is sown":** Theophrastus 1916.

17   **layers of the surrounding fruit:** The impenetrable shell of an *almendro* seed, for instance, consists primarily of *endocarp*, the innermost layer of the fruit.

17   **major divisions in the plant kingdom:** It speaks to the fundamental importance of seeds in plant evolution that they are used to define so many lineages: *gymnosperms* (the "naked seeds"); *angiosperms* (the flowering plants, or "enclosed seeds"); *monocots* (angiosperms with one cotyledon); *dicots* (angiosperms with two cotyledons). Even fine-scale relationships among species and closely related groups can often be made based on the structure of their seeds.

Chapter Two: The Staff of Life

20   **grass pride in a grass place:** One of the unexpected results of this work was the rediscovery of the giant Palouse earthworm (*Driloleirus americanus*), a native species long thought to have gone extinct with the decline of the prairies. Though recent specimens are smaller, this albino night crawler is rumored to reach three feet in length and to smell distinctly of lilies!

22   **70 percent of the land in cultivation:** Traditionally, the word *cereal* described the edible seeds of annual grasses, while *grain*

was a more general term that included the seeds of plants like buckwheat (in the same family as rhubarb) or quinoa (related to beets and spinach). As a result of the overwhelming success of the W. K. Kellogg and C. W. Post companies, however, cereal has now become inextricably linked with breakfast food, leaving grain as a catchall for grass and grass-like crops. It's too bad, since cereal was more descriptive and took its root from the lovely Ceres, Roman goddess of agriculture.

22 **grass seeds still feed the world:** Technically, each "grain" of a grass is a tiny fruit called a *caryopsis*. The fruit layers have adapted to serve as a hardened seed coat, however, and are indistinguishable from the seed material even under magnification. In all but the strictest interpretations, the caryopsis can be considered a de facto seed.

23 **unoccupied real estate:** Grasses evolved during an arid period in the Early Eocene, with a suite of traits adapted to life on the open plains. They are wind-pollinated and grow from the base, low to the ground, which helps them recover quickly from grazing or wildfires. Their leaves even contain glass-like silica crystals designed to wear down the teeth of the bison, horses, and other herd animals that feed on them.

24 **successful strategy:** It's also worth keeping in mind that while grass seeds appear small to us, they can be quite large compared to the size of the plant and do represent a considerable investment of energy, particularly for annuals.

25 **like a steel wire:** For a very enjoyable account of the chemistry behind this statement, see Chapter 4 in Le Couteur and Burreson's book *Napoleon's Buttons* (2003).

26 **" . . . the cooking ape":** According to Wrangham, modern proponents of a raw-food diet survive only through the largess of well-stocked grocery stores, and even then show signs of nutritional stress. In a natural setting, where food resources are scattered and seasonal, people would starve without the substantial energetic boost that comes through cooking (see Wrangham 2009).

27 **fruit, nuts, and seeds:** In addition to cooking, mastery of fire gave ancient people the ability to gather honey by smoking bees from their nests. For a fascinating discussion of this development, and our coevolution with a bird called the Greater Honeyguide, see Wrangham 2011.

27   **provide an invaluable comparison:** For more information on the
     many anthropological and archaeological notes in this paragraph,
     see Clarke 2007; Reddy 2009; Cowan 1978; Piperno et al. 2004;
     Mercader 2009; and Goren-Inbar et al. 2004.

28   **before our species even evolved:** I am consciously using the in-
     clusive *sensu lato* definition of *H. erectus*, though some authors
     prefer to split this species into an earlier African form (*H. ergas-
     ter*) and a later Asian form (*H. erectus*). Isotopic evidence, as well
     as teeth wear on fossils, suggest that grass in the diet stretches
     back much farther, to early hominins like *Australopithecus*. Lee-
     Thorp et al. (2012) believe they were eating the fibrous roots of
     wetland grasses and sedges year round, in which case it's likely the
     more nutritious seeds would have been relished when in season.

28   **varieties of wheat, rye, and barley:** In addition to efficiency and
     convenience in planting, there is evidence that a sudden cool and
     dry climate shift helped push these early agriculturalists toward a
     few hearty varieties (Hillman et al. 2001).

28   **resources into seed production:** Advocates for sustainable ag-
     riculture have begun developing large-seeded perennial grasses
     as alternatives to annual grains. If successful, these crops offer
     considerable advantages in erosion control, carbon sequestration,
     and reduced dependence on fertilizers and herbicide (Glover
     et al. 2010).

29   **" . . . explaining the course of human history":** For more on this
     statement, see Diamond 1999, 139, and Blumler 1998.

30   **" . . . destroyed the Western Empire":** Fraser and Rimas 2010,
     64.

30   **fur of the common black rat:** Symptoms of bubonic plague are
     brought on by *Yersinia pestis*, a bacteria passed among people and
     rats through flea bites. Though infected fleas also eventually suc-
     cumb, they can survive for many weeks as the bacteria population
     builds up in their mid-gut.

35   **" . . . bored shitless":** Harden 1996, 32.

35   **$4 billion in today's currency:** For more information on the his-
     tory of the Snake River dams, see Peterson and Reed 1994 and
     Harden 1996.

37   **both in flavor and nutrition:** "Complete" proteins contain us-
     able amounts of nine amino acids that the body needs yet can't
     produce itself. They must be obtained through diet. Most meat

and dairy proteins are complete, but many plant foods lack one or more of the essential amino acids.

CHAPTER THREE:
SOMETIMES YOU FEEL LIKE A NUT

40 **replacement for cane sugar:** Corn syrup comes directly from the starch found in maize seeds and can be purchased in the baking aisle of any grocery store. It is distinct from high fructose corn syrup, a variety processed with enzymes to intensify its sweetness. For an entertaining and informative look at the issues surrounding high fructose corn syrup, and the corn industry in general, I highly recommend the 2008 documentary *King Corn.*

41 **pressing the beans for butter:** Until the nineteenth century, people enjoyed chocolate primarily as a beverage, and its butter content was seen as an oily nuisance. The "Dutching" process (perfected by Holland's van Houten family) removed butter from the nibs with the aim of making a better drinking chocolate. Only later did chocolatiers mix that extra fat back in with whole ground beans to invent the modern chocolate bar. For more on the fascinating history and science of chocolate, see Beckett 2008 and Coe and Coe 2007.

42 *acellular endosperm:* Though ungainly on a product label, this term is very literally accurate. The coconut's liquid endosperm develops *without* defined cells—it is just a bunch of nuclei splashing around in a puddle of cytoplasm! Other seed endosperms may experience a free nuclei stage early in their development, but only coconuts maintain this bizarre arrangement at maturity.

43 **perhaps thousands of miles:** In spite of the coconut's remarkable dispersal ecology, its best trick for moving so far and wide has been to make itself useful to people. Virtually every native culture in the coastal tropics relies on coconuts in some way, and they've brought them along with them wherever they've traveled. Some evidence points to a Southeast Asian origin, but they were widespread from the South Pacific to Africa and South America long before botanists started asking questions.

44 **and then around the world:** While almond cultivation is widespread, it truly took root in California's Central Valley, where several thousand orchards now produce over 80 percent of the world's annual harvest. Nearly all California growers belong

to the Blue Diamond Cooperative, whose savvy marketing has helped almonds surpass grapes to become the state's most valuable crop.

45 **product called "rape oil":** Crop researchers at the University of Manitoba bred the current strain of commercial canola mustards to produce a more palatable, low-acid oil. The name comes from "Canadian oil, low acid."

48 **fine musical instruments:** Until the advent of cheap plastics after World War II, sliced and polished tagua nuts made up as much as 20 percent of the button market in North America and Europe. They've recently been making a comeback in the fashion industry. See Acosta-Solis 1948 and Barfod 1989 for more on the history of this beautiful seed.

48 **"brain worm":** In his 2008 book *Musicophilia*, Sacks notes an earlier and even more descriptive term used in Scotland for any maddeningly catchy tune: "the piper's maggot."

48 **PGPR . . . , from castor beans:** PGPR also contains glycerol, and the seed fat component sometimes comes from soybeans.

50 **eight times as effective as starch:** Though best known as a thickener, guar means something else entirely to firefighters, pipeline operators, or the designers of ship hulls and torpedoes. In minute quantities, it has the ability to create "slippery water," a phenomenon that greatly reduces drag. One physicist described molecules of guar gum (and similar polymers) as double yo-yos, coiling and uncoiling in such a way that they prevent turbulent liquids from adhering to adjacent surfaces. The physics are still poorly understood, but in practice this effect speeds the movement of fluids through hoses and pipes. The US Navy has also studied it as a way to increase hull efficiency and reduce the noise of its ships, submarines, and torpedoes.

52 **"the Pennsylvanian" in their honor:** American geologists once considered the Pennsylvanian to be its own, full-fledged period, but it's now considered a subdivision of the Carboniferous.

CHAPTER FOUR: WHAT THE SPIKE MOSS KNOWS

64 **by the early Carboniferous:** Clear precursors to the seed habit appeared in the late Devonian, including primitive seed ferns with ovule-like structures and *Archaeopteris*, an ancient tree that was one of the first woody plants with male and female spores.

66 **precursors to eggs:** Technically, these large spores evolved into what botanists call *ovules*, reproductive structures that include eggs and several layers of surrounding tissue.

66 **step in the evolution of seeds:** While it's easy to dismiss spike mosses and other modern spore plants as a minor part of the flora, they remain quite successful. Though the spore strategy is no longer dominant, it has persisted for hundreds of millions of years, and some lineages, particularly ferns, are more diverse now than ever.

67 **surface of leaves or cone scales:** Fruit-like tissues of various gymnosperms can be part of the seed itself (e.g., the red, berry-like aril of yew) or may derive from surrounding scales (e.g., the berry of a juniper). While they may carry out the same dispersal functions, these are not considered true fruits because they derive from different tissues.

68 **measured, incremental change:** See Friedman 2009.

69 **"the flowering plants":** No seed book would be complete without a rant against this annoying and misleading phrase! Yes, angiosperms have flowers and fruits, but so do many gymnosperms, both extant and extinct. Just as their names imply, it is a seed trait—the presence or absence of a carpel—that defines these important groups.

69 **" . . . experiments in coevolution":** Pollan 2001, 186.

CHAPTER FIVE: MENDEL'S SPORES

71 **in the spring of 1856:** Though he began his hybridization research in 1856, Mendel had devoted the previous two years to testing thirty-four local pea varieties, making sure they would breed true. In the end, he chose twenty-two of the most reliable strains for his experiments.

71 **free-swimming sperm:** Recent research shows that tiny mites and springtails may help transport moss sperm and facilitate fertilization (see Rosentiel et al. 2012). No one yet knows why they do it, but it's a fascinating reminder that we still have a lot to learn about the spore-plant reproductive system.

71 **anything else from the spores themselves:** One intriguing exception to this statement comes from nardoo, a type of floating fern in Australia that features male and female spores like a spike moss. The larger, female spores come in little packets that can be ground, washed, and baked into cakes. Though they taste nasty

and are quite toxic unless properly prepared, nardoo cakes once served as an important desperation food for several Aboriginal tribes. Famed Australian explorer Robert O'Hara Burke and several of his companions reportedly died after eating improperly cooked nardoo (see Clarke 2007).

72 **some were wrinkled, and some smooth:** In total, Mendel tracked the fate of seven seed and plant characteristics: wrinkled versus smooth, seed color, seed-coat color, pod shape, pod color, flower position, and stem length. To keep things simple, I focused on the first and most famous trait, wrinkled and smooth seeds.

72 **what he actually thought:** With so little original material to work from, most biographical sketches of Mendel include a fair amount of speculation. Iltis (1924 in German; 1932 in English) remains the main reference. It's an openly admiring portrait, but it benefits from the author's interviews with people who actually knew the monk.

73 **unopened and unread:** While it makes a wonderful story to say that Mendel's paper languished unread in Darwin's library, such accounts are patently false. Meticulous searches of Darwin's well-preserved collection have never uncovered a copy. Nor did he ever refer to Mendel's work in his writing or correspondence. The two did come within twenty miles of one another when Mendel visited the Great London Exhibition in 1862, but Darwin was home at Down House at the time, and there is no reason to believe they ever met.

75 **dramatically through selective breeding:** While most crops and local varieties developed incrementally over long time periods, the sophistication and pace of plant breeding increased rapidly during the Enlightenment in the seventeenth and eighteenth centuries. See Kingsbury (2009) for a good discussion of this history.

75 **protection from cross-pollination:** The qualities that made it so good for wine depended on maintaining a double-recessive gene, and would disappear if the millet interbred with other varieties. This "glutinous" or "waxy" mutation occurs in many edible grasses, including rice, sorghum, corn, wheat, and barley. It is always a recessive trait, but the results are sometimes considered a delicacy (e.g., moshi, botan, and other sticky rice varieties).

76 **at random from each parent:** Mendel's contributions to genetics are often summarized as the Law of Segregation (paired alleles, one from each parent) and the Law of Independent Assortment

(alleles passed down independently). He also gave us the terms *dominant* and *recessive*.

78 **bother with pollination at all:** The term *apomixis* describes several types of asexual reproduction in plants. For hawkweeds, dandelions, and many other members of the aster family, incomplete meiosis in the egg-forming process creates viable seeds that are essentially clones of the mother plant. Apomictic species may lose the advantages of regular genetic mixing, but they gain the ability to reproduce at will without reliance on pollinators (and most, if not all, retain the ability to reproduce normally in a pinch). If a species is well adapted, this strategy can be quite successful, as anyone with a lawn full of dandelions can attest to.

78 **" . . . changed the subject":** C. W. Eichling, as quoted in Dodson 1955.

79 **" . . . evolution of *particular* forms":** Bateson 1899. William Bateson, an eminent British botanist, made this statement in a talk to the Royal Horticultural Society. Quoted in more detail, his comments seem almost eerily prescient of Mendel's pending rediscovery: "What we first require is to know what happens when a variety is crossed with its *nearest allies*. If the result is to have a scientific value, it is almost absolutely necessary that the offspring of such crossing should then be examined *statistically*." Bateson went on to play a key role in championing Mendel's ideas, and he coined the term "genetics."

80 **three-to-one ratio:** The following year I planted my crosses and harvested a total of 1,218 peas. The ratio of smooth to wrinkled came in at 2.45 to 1, a close but not exact replication of Mendel's famous result. The difference might have been due to my smaller sample size, or perhaps to pollen contamination from the pea varieties in Eliza's nearby garden.

CHAPTER SIX: METHUSELAH

85 **endurance to the Jewish people:** For a fascinating examination of how the events at Masada transitioned from a historical footnote to a powerful story of heroism, see Ben-Yehuda 1995.

85 **set the building ablaze:** According to the Roman historian Josephus, the Sicarii left some portion of their provisions intact to show that they were well supplied until the end. This may explain

why some date seeds recovered from Masada are charred, while others are unburned.

85 **Jewish numismatics:** Until the finds at Masada, the provenance of certain coins minted during the Great Revolt had been considered "one of the most difficult problems of Jewish numismatics" (see Kadman 1957 and Yadin 1966).

86 **climate and settlement patterns:** After the Romans put down the Great Revolt and a subsequent uprising several decades later, the former kingdom of Judaea went into a steep decline. The export economy collapsed, whole towns were abandoned, and changing climate patterns made even small-scale date cultivation challenging. The once famous palm variety eventually winked out completely, as English clergyman and explorer Henry Baker Tristram wistfully observed on a visit in 1865: "The last palm has gone, and its graceful feathery crown waves no more over the plain, which once gave to Jericho its name of the City of Palm Trees" (Tristram 1865).

88 **example of a naturally germinating seed:** Elaine presoaked the seed with plant hormones and enzymatic fertilizers—standard techniques for germinating fragile samples—but the impetus to sprout was Methuselah's alone.

88 **flourish in the Jordan Valley:** Modern Israeli date stocks descend from standard cultivars imported in the twentieth century. Genetic testing shows that Methuselah is not related to any of these—he most closely resembles an old Egyptian variety called *hayani*. Though probably coincidental, this fits nicely with the traditional story that the Jews brought dates with them during the Exodus.

CHAPTER SEVEN: TAKE IT TO THE BANK

100 **extracted from rice and nuts:** Working with a $20 million grant from the Bill & Melinda Gates Foundation, a team led by Dr. Robert Sievers has developed live measles vaccines that remain viable for up to four years suspended in a "bioglass" of myo-insitol.

103 **" . . . going to survive."** Cary Fowler, quoted on *60 Minutes* (archived at "A Visit to the Doomsday Vault," March 20, 2008, CBS News, www.cbsnews.com/8301–18560_162–3954557.html).

104   **managed by Kew Gardens:** The Millennium Seed Bank cur-
      rently houses more than 2 billion seeds from over 34,000 different
      species, including more than 90 percent of the United Kingdom's
      native seed plant species. By 2025, the project aims to preserve
      seeds from 25 percent of the world's flora, with a focus on rare and
      threatened plants. Already, at least twelve species housed in the
      collection have gone extinct in the wild.

104   " . . . **time to begin is now.**" As quoted in Dunn 1944.

104   **resisting pests and disease:** Vavilov not only understood variety
      in cultivars, but also identified what he called "centers of origin,"
      eight regions in the world where important crops were originally
      domesticated, where they remain most diverse, and where their
      wild relatives can still be found. The idea remains an important
      principle in plant breeding and botanical research.

105   **promised quicker results:** Led by Trofim Lysenko, this notorious
      movement countered Mendelian genetics with a half-baked the-
      ory of environmentally acquired inheritance, setting back Soviet
      agriculture—and biology in general—for a generation.

105   **precious grains in their care:** Though Vavilov's collections sur-
      vived Lysenkoism and the destruction of World War II, the in-
      stitute that houses them has suffered funding setbacks and a long
      decline in the modern era. Its irreplaceable orchard—with more
      than 5,000 fruit and berry varieties—was recently slated to be
      cleared for a housing development.

106   **grown on a massive scale:** Similarly, the crisis in wild plant di-
      versity is also a result of human activities, from habitat loss to
      climate change to the introduction of invasive species.

Chapter Eight: By Tooth, Beak, and Gnaw

114   **gnawing seeds:** While the modern rodent diet includes a wide
      range of plant matter (and occasionally insects or meat), rodents'
      distinctive teeth evolved for gnawing seeds, which remain the
      most common food throughout the group.

117   **nut tucked safely inside:** Functionally, the pits of these species
      are seeds, but technically the shell is made up of a hardened fruit
      layer called an *endocarp*.

121   **shade of their parents:** There is a whole branch of dispersal ecol-
      ogy devoted to this idea. Getting away helps the young plant

avoid predation, competition with its parent and siblings, and the host of species-specific viruses and other pathogens that lurk in the vicinity of adult trees.

124 **changes in the finches:** Now entering its fifth decade, the Galapagos finch study is the most in-depth examination of evolution ever undertaken in the field. Led by Princeton biologists Peter and Rosemary Grant, the research has helped reveal how natural selection and other factors (genetic, behavioral, and environmental) work together to create and maintain species. Weiner's *Beak of the Finch* (1995) and the Grants' own *How and Why Species Multiply* (2008) are both highly recommended.

CHAPTER NINE: THE RICHES OF TASTE

128 **" . . . All Hot! Pepper Pot!":** This rhyme traces its origins to Philadelphia, where soup sellers in the eighteenth century sang it out while hawking their distinctive, spicy stew. Traditional recipes for Philadelphia pepper-pot soup call for various meat scraps, from tripe to turtle, but they all agree on the seasoning—a large dose of black peppercorns.

132 **$2.5 billion over that time:** Total average annual return includes all cash and spice dividends, plus the appreciation in stock from the company's founding to its high of 539 guilders in 1648. For more about the extraordinary history of this company, see de Vries and van der Woude 1997.

133 **" . . . impostor or a fool":** Young 1906, 206.

133 **turn up in time:** Though it seems preposterous in hindsight, Columbus can't be faulted for thinking the "Spice Islands" might be hard to pinpoint. Until well into the eighteenth century, nutmeg trees grew on fewer than ten of Southeast Asia's 25,000 islands. Cloves were found on only five.

136 **tissue that surrounds the seeds:** The white tissue, called *placenta*, manufactures capsaicin and retains about 80 percent of it. Approximately 12 percent is transferred to the seeds, and the remainder goes into the fruit tissue, mostly at the tip, where a nibbling animal might encounter it before doing too much damage.

139 **Plants . . . are stationary:** This kind of blanket statement invites exceptions, and the plant kingdom has been happy to oblige with various examples of plant motion, from the snapping of

Venus flytraps to the cringing leaves of sensitive plants to the imperceptible amble of a walking fig. Still, after seeds disperse and sprout, the overwhelming pattern for plant life is to remain rooted and still.

140 " . . . **burning and inflammation":** Appendino 2008, 90.

CHAPTER TEN: THE CHEERIEST BEANS

144 " . . . **slip of a pink":** This phrase, from the translation in Ukers (1922), refers to a common method for propagating carnations and other members of the pink family. Vegetative shoots can be "slipped" easily from the main stem just above the leaf nodes.

145 **impossible to unravel:** Most modern accounts of de Clieu's story derive from the version in William Ukers's 1922 classic, *All About Coffee*. I had some of de Clieu's original correspondence and part of a nineteenth-century French history retranslated, and many of Ukers's details checked out. But I couldn't find confirmation of the pirate attack!

146 " . . . **Martinico's shore":** This poem first appeared in the volume *Poems for Children*, coauthored by Charles Lamb and his sister Mary. Based on differences in style, as well as various notes and letters, scholars attribute this particular verse to Charles.

146 **Martinique to Mexico to Brazil:** The descendants of de Clieu's seedling provided founding stock for plantations throughout the French West Indies, and probably in Central and South America, too. How far they spread remains unclear, but a popular story in Brazil traces at least part of that country's coffee stock to French Guiana—and to another tale of theft and seduction. According to legend, romance between a visiting Portuguese officer and the governor's wife resulted in an unusual parting gift. When he left for Brazil, she presented him with a fragrant bouquet of flowers. Sprigs and seeds of the colony's closely guarded coffee plants were tucked away inside.

146 **"uncoordinated writhing":** See Hollingsworth et al. 2002.

147 **original versions of Coke and Pepsi:** Though their formulas are tightly held trade secrets, both Coke and Pepsi entered the soda market at a time when "colas" inherently included kola nut extract. Whether the modern versions still do remains a matter of debate, but a recent chemical analysis found no trace of kola-nut proteins in a can of regular Coke (D'Amato et al. 2011).

147 **the hardiest attackers:** Caffeine is considered an excellent all-around pesticide, but specialist insects like the coffee-borer beetle have developed an immunity. They have no trouble chomping their way through coffee beans and can cause extensive crop damage.

148 **other seeds from germinating:** Exactly how caffeine gets from the seed to the soil remains unclear—it may diffuse directly, or even pass through the root. And in a practical last stage of the plant's alkaloid recycling program, some caffeine moves from the endosperm to the seed leaves, protecting them from attackers and starting the whole process over again.

149 **dosed with caffeine:** The amount of caffeine in coffee nectar strongly suggests coevolution with honeybees. Too much is a bitter, even toxic deterrent, but coffee flowers provide just the right dose to stimulate memory and keep the bees coming back for more.

149 **" . . . expression rules for a short time":** from the *British Homeopathic Review*, as quoted in Ukers 1922, 175.

150 **Industrial Revolution that followed:** Experts have a charming name for this period of rapidly evolving work habits: the Industrious Revolution.

151 **300 to 400 liters considered average:** Annual totals as high as 1,095 liters per person have been reported from hospitals, where presumably beer was a cost-effective way of feeding the patients. See Unger 2004 for an excellent account of beer habits from the Middle Ages through the Renaissance.

152 **" . . . spiritually and ideologically":** Schivelbusch 1992, 39.

153 **" . . . goods of pirates":** The British Admiralty confiscated a trove of jewels, precious metals, and trade goods with the capture of Captain William Kidd in 1699. These items were later auctioned at London's Marine Coffee House, raising enough capital to establish a retirement facility for destitute sailors (see Zacks 2002, 399–401).

153 **Newton dissecting a dolphin:** It's an extremely appealing image, but this often-repeated yarn can be easily debunked by Ralph Thoresby's eyewitness account, which tells of retiring to The Grecian *after* the dissection (Thoresby 1830, vol. 2, 117). What is perhaps more interesting is that the dolphin was caught nearby in the River Thames!

154 **Franklin dropped by:** Perhaps nothing demonstrates Franklin's legendary popularity in France better than the reaction at Café

Procope upon news of his death. For three days of mourning the inner room was hung with black fabric. Memorial speeches were read, and patrons decorated a bust of Franklin with a crown of oak leaves, cypress boughs, celestial charts, globes, and a serpent biting its tail—a sign of immortality.

Chapter Eleven: Death by Umbrella

164 **arrested for the crime:** Documentarian Richard Cummings believes that Markov's murder involved a team of assassins, including a getaway driver at the wheel of a taxi cab. In this version of the story, the dropped umbrella was meant only as a distraction, while a small, pen-sized object actually delivered the deadly pellet.

165 **cell's genetic code into action:** The chain released inside a cell interferes with RNA transcription, essentially stopping the cell's ability to synthesize the proteins that make it function. On its own, unable to penetrate cells, this chain is quite harmless and closely resembles the storage proteins in a range of commonly eaten seeds, including barley.

167 **description of the symptoms:** Final confirmation of ricin as the poison came from the failed assassination attempt in Paris. Because the full dose failed to disperse, the intended victim survived. His body did, however, produce antibodies to the trace amount of ricin that entered his bloodstream.

167 **put the plan into action:** Kalugin 2009, 207.

169 **working in labs around the world:** See Preedy et al. 2011.

171 **kill a full-grown cow:** This fact would not surprise mycologists like Noelle Machnicki one bit. Research continues to show that many botanical compounds are actually the product of plant-fungi interactions, and in some cases they are made entirely by fungi living on or within the plant.

Chapter Twelve: Irresistible Flesh

180 **ever laying eyes on one:** The main long-distance disperser of *almendro* is the great fruit-eating bat (*Artibeus lituratus*). The Jamaican fruit-eating bat occasionally disperses the fruit as well, but other bat species are considered too small to carry a normal-sized *almendro* fruit (see Bonaccorso et al. 1980).

181 **helped me take the data further:** Another aspect of my research tracked the dispersal of pollen and found a similar result. Bees lured by *almendro*'s prolific purple flowers would fly nearly a mile and a half (2.3 kilometers) between trees, moving pollen among even the most isolated individuals.

184 **attract otherwise carnivorous ants:** Called *eliasomes*, these fatty, protein-rich blobs lie at the heart of ant-plant interactions. The strategy has evolved at least one hundred times in groups as distinct as sedges, violets, and acacia trees. While most ant-dispersal distances are short, seeds in at least one case have been carried nearly 600 feet (180 meters) (Whitney 2002).

185 **" . . . out of their minds":** Cohen 1969, 132.

185 **(and as yet unknown) seed disperser:** Many botanists include *manzanillo* on the list of plants possibly dispersed by gomphotheres or some other long-extinct megafauna.

188 **common in gymnosperms:** While people associate fruits with the flowering plants, animal dispersal is actually far more widespread in the gymnosperms. It occurs in 64 percent of gymnosperm families, and only 27 percent of angiosperm families (see Herrera and Pellmyr 2002 and Tiffney 2004).

190 **particular types of dispersers:** Botanists call these strategies *dispersal syndromes*. But while they offer a useful way to categorize plant-animal interactions, their role in actually driving plant evolution remains controversial.

190 **fertilizing dung:** In some cases the dung pile is beneficial, but when it contains too many seeds that sprout all at once, the intense competition can counteract the benefits of the fertilizer.

Chapter Thirteen: By Wind and Wave

193 **" . . . Daisy from a Dandelion":** From letter to J. D. Hooker, 1846 (van Wyhe 2002).

193 **" . . . no botanist":** Letter from J. S. Henslow to W. J. Hooker, 1836, as cited in Porter 1980.

194 **"Brazil without big trees":** from Darwin's Galapagos Notebook (van Wyhe 2002).

194 **" . . . arctic, than an equatorial Flora":** Darwin 1871, 374.

195 **" . . . their nature but nothing else":** Columbus 1990, 97.

195 **" . . . fields like roses":** Cohen 1969, 79.

197 **idea took hold:** Mandeville's original text describes the gourd
     lambs as "without wool" and suggests they were grown for the
     table rather than the loom. He even claims to have tasted one,
     proclaiming the flavor "wonderful." Recent interpretations often
     omit this text, however. The Mandeville quotation about cotton
     trees with "pliable branches" and "hungrie" lambs appears to be a
     complete fabrication, spread in part by its inclusion in his Wiki-
     pedia entry.

200 **375 feet (120 meters):** Dauer et al. 2009.

200 **prey upon the springtails:** See Swan 1992 for a fascinating de-
     scription of this wind-dependent high-altitude ecosystem, which
     he calls "the Aeolian biome."

200 **300 miles (483 kilometers) on an average Atlantic current:** Dar-
     win later upped that estimate to 924 miles (1,487 kilometers) when
     he realized that whole, dried plants stayed afloat much longer.

201 **" . . . coming to maturity!":** Darwin 1859, 228.

201 **archipelago by similar means:** Porter (1984) attributed 134 col-
     onizations to wind and 36 to drift, allowing that some, like cot-
     ton, were a combination of both.

201 **"a miracle squared":** de Queiroz 2014, 287.

202 **"the revolutionary fiber," and "the fuel of the industrial revo-
     lution":** Yafa 2005, 70; Riello 2013, 2.

202 **" . . . you have no modern industry":** McLellan 2000, 221.

203 **Europe's growing middle class:** Simply reading the tags at a
     modern fabric store quickly reveals the roles of Asia and the Near
     East in the history of cotton. In addition to calicos from Cali-
     cut, one finds madras (from the city of Madras), chintzes (from
     the Hindi word for "paint" or "splatter"), khakis (from the Urdu
     for "dust-colored"), gingham (from the Malay word for "striped"),
     and seersucker (from the Persian for "milk and sugar," a reference
     to the bunched and smooth pattern of the fabric).

204 **its African and Asian relations:** The long-fibered New World
     cottons resulted from hybridization between two Old World va-
     rieties. They are what geneticists call *tetraploids*, with double the
     normal number of chromosomes. Of the five recognized species,
     upland cotton (G. *hirsutum*) now dominates the world market.
     Sea Island cotton (G. *barbadense*) has the longest fibers, but it is
     harder to grow. It persists in the marketplace for high-end fabrics,
     usually under the trade names "Egyptian Cotton" or "Peruvian
     Pima."

205 **Middle Passage to America every year:** See Klein 2002.

207 **twice that rate:** The winged samaras of a maple may fall faster than a Javan cucumber seed, but they have inspired aircraft of their own. Lockheed Martin manufactures the "Samarai," a surveillance drone that gyrates like a maple, and Australian researchers recently unveiled a disposable whirligig designed to transmit atmospheric conditions from the air above a forest fire. Single-rotor helicopters have also been built, but they generally lack the stability necessary for manned flight.

210 **laughed until it disappeared from view:** Watching that Javan cucumber seed fly was a thrill, but I also felt a certain unease as it drifted out of sight. What if it sprouted? Though it's extremely unlikely that a tropical vine would thrive in our cool climate, I couldn't help wondering whether Noah and I had just introduced the future kudzu of the Pacific Northwest!

Conclusion: The Future of Seeds

212 **doubled chemically:** The chemical in question is *colchicine*, an alkaloid found in the seeds and tubers of the autumn crocus.

Glossary

acellular endosperm An unusual substance found in coconuts, known in the grocery store as "coconut water." It consists of free nuclei floating in a nutritious cytoplasmic bath. As the coconut matures, cell walls form, and much of this material transforms into the meat (i.e., solid endosperm) of the coconut. (Some other seed endosperms pass through a brief acellular stage very early in their development, but only coconuts maintain it for so long and in such quantities.)

adenosine A compound with a wide range of vital functions in biochemistry. In the brain, it plays an important role in signaling fatigue and guiding the body toward sleep.

alkaloid Any of a large group of nitrogen-based compounds produced by plants and some marine organisms. They often function as chemical defenses, and many of them, including stimulants (e.g., caffeine), drugs (e.g., morphine), and poisons (e.g., strychnine), produce a strong reaction in people.

allele One of the possible forms of a gene, determined by differences in DNA and resulting in different expressions of that gene (e.g., wrinkled vs. smooth peas, or brown vs. red hair in people).

angiosperm A "flowering plant," defined by having its seed enclosed in tissue to form a carpel (see *carpel*, below). The vast majority of living plants are angiosperms.

apomixis Asexual reproduction in plants that occurs when egg cells are produced with a full set of chromosomes, requiring no fertilization by pollen. The resulting seeds are essentially clone-like offspring of the parent. This strategy has evolved occasionally in a wide range of plant families, but is perhaps most common in the

245

asters, including the dandelion, and the hawkweeds that so confounded Gregor Mendel.

caffeine   An alkaloid found in a number of plants (notably coffee, tea, kola, and cacao) that helps deter attacks by insects and other pests, and may also function in the soil as an herbicide and germination inhibitor. Used by people as a stimulant.

carbohydrate   A group of compounds in biochemistry composed of carbon, hydrogen, and oxygen atoms in various combinations. The results are generically called *sugars*, but they can be used for everything from energy storage in seeds (e.g., starch) to the exoskeleton of insects (called *chitin*).

Carboniferous   The fifth Period of the Paleozoic Era, following the Devonian and lasting from 360 million to 286 million years ago (includes the Mississippian and Pennsylvanian sub-periods).

carpel   The defining characteristic of the angiosperms that evolved from the leaves or bracts that surrounded and enclosed the seed, forming a protective layer and spurring a host of new adaptations for defense, pollination, and dispersal. One or more carpels make up what is considered the female portion of a typical angiosperm flower, including the ovary, stigma, and style.

caryopsis   A type of fruit generally understood as the "seeds" of grasses.

cereal   An annual, grain-bearing grass (e.g., wheat, barley, rye, oats, corn, rice).

chromosome   The structure that bears the genetic information of a plant or animal. Chromosomes consist of the double-helix DNA molecule and surrounding proteins and serve as bulk units of inheritance between generations. In sexual reproduction, individuals receive half their chromosomes from each parent.

coevolution   Process of evolution where changes in one organism spur changes in another. Traditionally, coevolution has been defined as a reciprocal interaction between two species, but it is now understood to be far more nuanced, producing changes within networks of interacting species that can vary across geography and through time.

copra   The "meat" of a coconut, formed from solid, cellular endosperm.

cotyledon   The embryonic leaf of a baby plant, also called a "seed leaf." Cotyledons are well known to gardeners as the first leaves of germinating seedlings, and are also familiar when they are particularly large and tasty within the seed itself (e.g., the two halves of a peanut).

Cretaceous The final period of the Mesozoic Era, following the Jurassic and lasting from 146 million to 65 million years ago.

cytotoxin A poison that physically kills individual cells, as opposed to a neurotoxin, which causes paralysis or other damage to the nervous system.

dicot A major group of flowering plants defined by the presence of two cotyledons (*di-cot*) in the seed.

diploid The condition of having two sets of chromosomes, one from each parent.

dormancy Generally understood as the period of inactivity between the maturation of a seed and its germination. Technically, true dormancy applies only to those seeds that actively resist germination until various physical or chemical requirements are met (e.g., changes in light, temperature, and moisture, or exposure to wood smoke).

electron micrograph An image taken at extreme magnification by an electron microscope.

eliasome A rich, fatty packet attached to a seed to encourage seed dispersal by ants.

embryo In general, an unborn offspring. In botany, this term refers to the baby plant found within a seed.

emulsifier A substance (e.g., lecithin) added to stabilize the suspension of one liquid within another. In food products, emulsions are usually oils or fats suspended in water (e.g., mayonnaise), but can also be water in a fat (e.g., butter). Emulsifiers can also help suspend particles in a liquid, as with sugar and chocolate solids in cocoa butter.

endocarp The innermost layer of a fruit, often hardened to protect the seed.

endorphin One of a group of hormones secreted by the central nervous system. Endorphins are generally believed to be involved in regulating pain and pleasure responses.

endosperm An important tissue for the storage of nutrition in seeds. In angiosperms, it is technically a triploid product of pollination. In gymnosperms, this role is played by the megagametophyte.

endozoochory Literally, "going abroad within animals." The term refers to the seed dispersal strategy of being consumed, transported, and deposited by an animal.

enzyme Compounds, usually proteins, produced to catalyze a chemical reaction within an organism.

epicotyl   Literally, "above the leaf." It refers to the stem-like portion of a baby plant that is above the seed leaves and below the shoot, or *plumule*.

gametophyte   Literally, "gamete-producing plant." It refers to an independent generation in the spore-plant life cycle that produces eggs and sperm. In ferns, for example, it is a tiny, separate plant that grows from a spore and lives briefly in damp soil.

gene   A specific location along a chromosome, where the shape and pattern of the DNA determine a specific trait.

genetically modified organism (GMO)   A plant, animal, or microbe whose genetic code has been artificially altered, typically by the deletion or manipulation of genes, or by inserting genes from another organism.

germination   The awakening of a seed. Technically, this process begins with water uptake (see *imbibition*) and ends when the radicle emerges from the seed coat. More generally, it includes the full emergence and establishment of the baby plant's root and shoot.

grain   Cereals (e.g., wheat, rice) and other similar crops (e.g., quinoa, buckwheat).

gymnosperm   Literally, "naked seed." The gymnosperms are a major group of seed plants defined by the lack of a carpel or enclosure around the seed.

hormone   Any of a range of compounds that regulate growth, development, and other processes within a plant or animal.

hybrid   A cross between two species, or between two distinct varieties of the same species.

hypocotyl   Literally, "below the leaf." The term refers to the stem-like portion of a baby plant that is beneath the seed leaves and above the root, or *radicle*.

imbibition   The rapid uptake of water by a seed that signals the beginning of germination.

in situ   Latin for "in place." This phrase is commonly used in conservation and the natural sciences to describe activities within the natural habitat of a species. (*Ex situ*, in contrast, describes the study or conservation of a species in a zoo or nursery setting).

kernel   A term for seed usually applied to cereals or to the soft, edible portion of tree nuts.

lecithin   A fatty substance extracted from the storage oils in certain seeds, including soybeans, rape seeds, cottonseeds, and sunflower

seeds. It is used as an emulsifier in food products and also as a cho-
lesterol-reducing dietary supplement.

megagametophyte   Literally, the "large-gamete-producing plant." The
term refers to egg-producing tissues that are a stand-alone plant in
old lineages like the spike moss, but are incorporated into the flowering parts of seed plants. The megagametophyte produces the egg,
and then its tissues are often included as part of the seed. Conifers
and other gymnosperms, for example, pack the energy (or "lunch")
for their seeds in the megagametophyte.

meiosis   Cell division that produces eggs and sperm or pollen. Instead
of typical division (*mitosis*), where all chromosomes are duplicated,
meiosis results in cells containing only half the normal chromosome allotment.

meristem   The parts of a plant where cell division takes place, typically found at the tips of roots and shoots, and also around the
perimeters of the stems and trunks of woody species.

metabolism   The sum of all chemical reactions and processes occurring within an organism, generally considered the basis of life.

monocot   A major group of flowering plants, defined by having one
cotyledon (*mono-cot*) in the seed.

paleobotany   The study of ancient plants.

Pennsylvanian   A sub-period of the Carboniferous Period, also re-
ferred to as the Upper Carboniferous, which lasted from 323 million to 290 million years ago.

perisperm   A starchy storage tissue found in seeds alongside (or, rarely,
in place of) the endosperm.

Permian   The sixth and final period of the Paleozoic Era, following
the Carboniferous and lasting from 290 million to 245 million
years ago.

photosynthesis   The use of sunlight to transform water and carbon
dioxide into life-sustaining carbohydrates, producing oxygen as a
by-product.

pip   A term for seeds generally applied to small, hard seeds found
within soft fruits.

plumule   The shoot of a plant embryo.

pulse   A term for the edible seeds of various leguminous crops, including beans, lentils, and chickpeas.

radiation   A rapid divergence and diversification of new species from
an ancestral form.

radicle   The root of a plant embryo.

recalcitrant   A seed that does not desiccate and lacks a truly quiescent or dormant stage.

ribosome   An organelle within cells that regulates the translation and expression of genetic information to produce proteins.

seed coat   The outermost layer of the true seed, often serving protective, waterproofing, or dispersal functions, and sometimes intermingled with surrounding fruit tissues.

spore   A tiny reproductive unit used by ferns, mosses, spike mosses, and other ancient plant groups. The seed habit evolved from the spore plants.

stamen   The "male" part of the flower, bearing pollen-producing anthers.

stigma   The area of the pistil, or "female" part of a flower, that receives the pollen.

tetraploid   The condition of having four sets of chromosomes, two from each parent.

Theophrastus   A student of and successor to Aristotle at the Lyceum. He is particularly well-known for his plant studies, and is often called "the father of botany."

triploid   The condition of having three sets of chromosomes, derived from hybridization between diploid and tetraploid parents.

Bibliography

Acosta-Solis, M. 1948. Tagua or vegetable ivory: A forest product of Ecuador. *Economic Botany* 2: 46–57.

Alperson-Afil, N., D. Richter, and N. Goren-Inbar. 2007. Phantom hearths and controlled use of fire at Gesher Benot Ya'aqov, Israel. *Paleoanthropology* 1: 1–15.

Alperson-Afil, N., G. Sharon, M. Kislev, Y. Melamed, et al. 2009. Spatial organization of hominin activities at Gesher Benot Ya'aqov, Israel. *Science* 326: 1677–1680.

Anaya, A. L., R. Cruz-Ortega, and G. R. Waller. 2006. Metabolism and ecology of purine alkaloids. *Frontiers in Bioscience* 11: 2354–2370.

Appendino, G. 2008. Capsaicin and Capsaicinoids. Pp. 73–109 in E. Fattoruso and O. Taglianatela-Scafati, eds., *Modern Alkaloids*. Weinheim: Wiley-VCH.

Asch, D. L., and N. B. Asch. 1978. The economic potential of *Iva annua* and its prehistoric importance in the Lower Illinois Valley. Pp. 300–341 in R. I. Ford, ed., *The Nature and Status of Ethnobotany*. Anthropological Papers No. 67. Ann Arbor: University of Michigan Museum of Anthropology.

Ashihara, H., H. Sano, and A. Crozier. 2008. Caffeine and related purine alkaloids: Biosynthesis, catabolism, function and genetic engineering. *Phytochemistry* 68: 841–856.

Ashtiania, F., and F. Sefidkonb. 2011. Tropane alkaloids of *Atropa belladonna* L. and *Atropa acuminata* Royle ex Miers plants. *Journal of Medicinal Plants Research* 5: 6515–6522.

Atwater, W. O. 1887. How food nourishes the body. *Century Illustrated* 34: 237–251.

———. 1887. The potential energy of food. *Century Illustrated* 34: 397–251.

Barfod, A. 1989. The rise and fall of the tagua industry. *Principes* 33: 181–190.

Barlow, N., ed. 1967. *Darwin and Helsow: The Growth of an Idea. Letters, 1831–1860.* London: John Murray.

Baskin, C. C., and J. M. Baskin. 2001. *Seeds: Ecology, Biogeography, and Evolution of Dormancy and Germination.* San Diego: Academic Press.

Bateman, R. M., P. R. Crane, W. A. DiMichele, P. Kenrick, et al. 1998. Early evolution of land plants: Phylogeny, physiology, and ecology of the primary terrestrial radiation. *Annual Review of Ecology and Systematics* 29: 263–292.

Bateson, W. 1899. Hybridisation and cross-breeding as a method of scientific investigation. *Journal of the Royal Horticultural Society* 24: 59–66.

———. 1925. Science in Russia. *Nature* 116: 681–683.

Baumann, T. W. 2006. Some thoughts on the physiology of caffeine in coffee—and a glimpse of metabolite profiling. *Brazilian Journal of Plant Physiology* 18: 243–251.

Bazzaz, F. A., N. R. Chiariello, P. D. Coley, and L. F. Pitelka. 1987. Allocating resources to reproduction and defense. *BioScience* 37: 58–67.

Beckett, S. T. 2008. *The Science of Chocolate*, 2nd ed. Cambridge, UK: Royal Society of Chemistry Publishing.

Benedictow, O. J. 2004. *The Black Death: The Complete History.* Woodbridge, UK: Boydell Press.

Ben-Yehuda, N. 1995. *The Masada Myth: Collective Memory and Mythmaking in Israel.* Madison: University of Wisconsin Press.

Berry, E. W. 1920. *Paleobotany.* Washington, DC: US Government Printing Office.

Bewley, J. D., and M. Black. 1985. *Seeds: Physiology of Development and Germination.* New York: Plenum Press.

———. 1994. *Seeds: Physiology of Development and Germination*, 2nd ed. New York: Plenum Press.

Billings, H. 2006. The *materia medica* of Sherlock Holmes. *Baker Street Journal* 55: 37–44.

Black, M. 2009. Darwin and seeds. *Seed Science Research* 19: 193–199.

Black, M., J. D. Bewley, and P. Halmer, eds. 2006. *The Encyclopedia of Seeds: Science, Technology, and Uses.* Oxfordshire, UK: CABI.

Blumler, M. 1998. Evolution of caryopsis gigantism and the origins of agriculture. *Research in Contemporary and Applied Geography: A Discussion Series* 22 (1–2): 1–46.

Bonaccorso, F. J., W. E. Glanz, and C. M. Sanford. 1980. Feeding assemblages of mammals at fruiting *Dipteryx panamensis* (Papilionaceae) trees in Panama: Seed predation, dispersal and parasitism. *Revista de Biología Tropical* 28: 61–72.

Browne, J., A. Tunnacliffe, and A. Burnell. 2002. Plant desiccation gene found in a nematode. *Nature* 416: 38.

Campos-Arceiz, A., and S. Blake. 2011. Megagardeners of the forest: The role of elephants in seed dispersal. *Acta Oecologica* 37: 542–553.

Carmody R. N., and R. W. Wrangham. 2009. The energetic significance of cooking. *Journal of Human Evolution* 57: 379–391.

Chandramohan, V., J. Sampson, I. Pastan, and D. Bigner. 2012. Toxin-based targeted therapy for malignant brain tumors. *Clinical and Developmental Immunology* 2012: 15 pp., doi:10.1155/2012/480429.

Charles, D. 2002. *Lords of the Harvest: Biotech, Big Money, and the Future of Food*. New York: Basic Books.

Chen H. F., P. L. Morrell, V. E. Ashworth, M. De La Cruz, et al. 2009. Tracing the geographic origins of major avocado cultivars. *Journal of Heredity* 100: 56–65.

Clarke, P. A. 2007. *Aboriginal People and Their Plants*. Dural Delivery Center, New South Wales: Rosenberg Publishing.

Coe, S. D., and M. D. Coe. 2007. *The True History of Chocolate*, rev. ed. London: Thames and Hudson.

Cohen, J. M., ed. 1969. *Christopher Columbus: The Four Voyages*. London: Penguin.

Columbus, C. 1990. *The Journal: Account of the First Voyage and Discovery of the Indies*. Rome: Istituto Poligrafico e Zecca Della Stato.

Corcos, A. F., and F. V. Monaghan. 1993. *Gregor Mendel's Experiments on Plant Hybrids: A Guided Study*. New Brunswick, NJ: Rutgers University Press.

Cordain, L. 1999. Cereal grains: Humanity's double-edged sword. Pp. 19–73 in A. P. Simopolous, ed., *Evolutionary Aspects of Nutrition and Health: Diet, Exercise, Genetics and Chronic Disease*. Basel: Karger.

Cordain, L., J. B. Miller, S. B. Eaton, N. Mann, et al. 2000. Plant-animal subsistence ratios and macronutrient energy estimations in worldwide hunter-gatherer diets. *American Journal of Clinical Nutrition* 71: 682–692.

Cowan, W. C. 1978. The prehistoric use and distribution of maygrass in eastern North America: Cultural and phytogeographical implications. Pp. 263–288 in R. I. Ford, ed., *The Nature and Status of Ethnobotany*. Anthropological Papers No. 67. Ann Arbor: University of Michigan Museum of Anthropology.

Crowe, J. H., F. A. Hoekstra, and L. M. Crowe. 1992. Anhydrobiosis. *Annual Review of Physiology* 54: 579–599.

Cummings, C. H. 2008. *Uncertain Peril: Genetic Engineering and the Future of Seeds*. Boston: Beacon Press.

D'Amato, A., E. Fasoli, A. V. Kravchuk, and P. G. Righetti. 2011. Going nuts for nuts? The trace proteome of a cola drink, as detected via combinatorial peptide ligand libraries. *Journal of Proteome Research* 10: 2684–2686.

Darwin, C. 1855. Does sea-water kill seeds? *The Gardeners' Chronicle* 21: 356–357.

———. 1855. Effect of salt water on the germination of seeds. *The Gardeners' Chronicle* 47: 773.

———. 1855. Effect of salt water on the germination of seeds. *The Gardeners' Chronicle* 48: 789.

———. 1855. Longevity of seeds. *The Gardeners' Chronicle* 52: 854.

———. 1855. Vitality of seeds. *The Gardeners' Chronicle* 46: 758.

———. 1856. On the action of sea-water on the germination of seeds. *Journal of the Proceedings of the Linnean Society of London, Botany* 1: 130–140.

———. 1859. *On the Origin of Species by Means of Natural Selection*. Reprint of 1859 first edition. Mineola, NY: Dover.

———. 1871. *The Voyage of the Beagle*. New York: D. Appleton.

Dauer, J. T., D. A. Morensen, E. C. Luschei, S. A. Isard, et al. 2009. *Conyza canadensis* seed ascent in the lower atmosphere. *Agricultural and Forest Meteorology* 149: 526–534.

Davis, M. 2002. *Dead Cities*. New York: New Press.

Daws, M. I., J. Davies, E. Vaes, R. van Gelder, et al. 2007. Two-hundred-year seed survival of *Leucospermum* and two other woody species from the Cape Floristic region, South Africa. *Seed Science Research* 17: 73–79.

DeJoode, D. R., and J. F. Wendel. 1992. Genetic diversity and origin of the Hawaiian Islands cotton, *Gossypium tomentosum*. *American Journal of Botany* 79: 1311–1319.

de Queiroz, A. 2014. *The Monkey's Voyage: How Improbable Journeys Shaped the History of Life*. New York: Basic Books.

de Vries, J. A. 1978. *Taube, Dove of War*. Temple City, CA: Historical Aviation Album.

De Vries, J., and A. van der Woude. 1997. *The First Modern Economy: Success, Failure, and Perseverance of the Dutch Economy, 1500–1815*. Cambridge, UK: Cambridge University Press.

Diamond, J. 1999. *Guns, Germs, and Steel: The Fate of Human Societies*. New York: W. W. Norton.

DiMichele, W. A., and R. M. Bateman. 2005. Evolution of land plant diversity: Major innovations and lineages through time. Pp. 3–14 in G. A. Krupnick and W. J. Kress, eds., *Plant Conservation: A Natural History Approach*. Chicago: University of Chicago Press.

DiMichele, W. A., J. I. Davis, and R. G. Olmstead. 1989. Origins of heterospory and the seed habit: The role of heterochrony. *Taxon* 38: 1–11.

Dodson, E. O. 1955. Mendel and the rediscovery of his work. *Scientific Monthly* 81: 187–195.

Dunn, L. C. 1944. Science in the U.S.S.R.: Soviet biology. *Science* 99: 65–67.

Dyer, A. F., and S. Lindsay. 1992. Soil spore banks of temperate ferns. *American Fern Journal* 82: 9–123.

Emsley, J. 2008. *Molecules of Murder: Criminal Molecules and Classic Cases*. Cambridge, UK: Royal Society of Chemistry.

Enders, M. S., and S. B. Vander Wall. 2012. Black bears *Ursus americanus* are effective seed dispersers, with a little help from their friends. *Oikos* 121: 589–596.

Evenari, M. 1981. The history of germination research and the lesson it contains for today. *Israel Journal of Botany* 29: 4–21.

Falcon-Lang, H., W. A. DiMichele, S. Elrick, and W. J. Nelson. 2009. Going underground: In search of Carboniferous coal forests. *Geology Today* 25: 181–184.

Falcon-Lang, H. J., W. J. Nelson, S. Elrick, C. V. Looy, et al. Incised channel fills containing conifers indicate that seasonally dry vegetation dominated Pennsylvanian tropical lowlands. *Geology* 37: 923–926.

Faust, M. 1994. The apple in paradise. *HortTechnology* 4: 338–343.

Finch-Savage, W. E., and G. Leubner-Metzger. 2006. Seed dormancy and the control of germination. *New Phytologist* 171: 501–523.

Fitter, R. S. R., and J. E. Lousley. 1953. *The Natural History of the City*. London: Corporation of London.

Fraser, E. D. G., and A. Rimas. 2010. *Empires of Food: Feast, Famine, and the Rise and Fall of Civilizations*. New York: Free Press.

Friedman, C. M. R., and M. J. Sumner. 2009. Maturation of the embryo, endosperm, and fruit of the dwarf mistletoe *Arceuthobium americanum* (Viscaceae). *International Journal of Plant Sciences* 170: 290–300.

Friedman, W. E. 2009. The meaning of Darwin's "abominable mystery." *American Journal of Botany* 96: 5–21.

Gadadhar, S., and A. A. Karande. 2013. Abrin immunotoxin: Targeted cytotoxicity and intracellular trafficking pathway. *PLoS ONE* 8: e58304. doi:10.1371/journal.pone.0058304.

Galindo-Tovar, M. E., N. Ogata-Aguilar, and A. M. Arzate-Fernández. 2008. Some aspects of avocado (*Persea americana* Mill.) diversity and domestication in Mesoamerica. *Genetic Resources and Crop Evolution* 55: 441–450.

Gardiner, J. E. 2013. *Bach: Music in the Castle of Heaven*. New York: Alfred A. Knopf.

Garnsey, P., and D. Rathbone. 1985. The background to the grain law of Gaius Gracchus. *Journal of Roman Studies* 75: 20–25.

Glade, M. J. 2010. Caffeine—not just a stimulant. *Nutrition* 26: 932–938.

Glover, J. D., J. P. Reganold, L. W. Bell, J. Borevitz, et al. 2010. Increased food and ecosystem security via perennial grains. *Science* 328: 1638–1639.

González-Di Pierro, A. M., J. Benítez-Malvido, M. Méndez-Toribio, I. Zermeño, et al. 2011. Effects of the physical environment and primate gut passage on the early establishment of *Ampelocera hottlei* (Standley) in rain forest fragments. *Biotropica* 43: 459–466.

Goor, A. 1967. The history of the date through the ages in the Holy Land. *Economic Botany* 21: 320–340.

Goren-Inbar, N., N. Alperson, M. E. Kislev, O. Simchoni, et al. 2004. Evidence of hominin control of fire at Gesher Benot Ya'aqov, Israel. *Science* 304: 725–727.

Goren-Inbar, N., G. Sharon, Y. Melamed, and M. Kislev. 2002. Nuts, nut cracking, and pitted stones at Gesher Benot Ya'aqov, Israel. *Proceedings of the National Academy of Sciences* 99: 2455–2460.

Gottlieb, O., M. Borin, and B. Bosisio. 1996. Trends of plant use by humans and nonhuman primates in Amazonia. *American Journal of Primatology* 40: 189–195.

Gould, R. A. 1969. Behaviour among the Western Desert Aborigines of Australia. *Oceania* 39: 253–274.

Grant, P. R., and B. R. Grant. 2008. *How and Why Species Multiply: The Radiation of Darwin's Finches*. Princeton, NJ: Princeton University Press.

Greene, R. A., and E. O. Foster. 1933. The liquid wax of seeds of *Simmondsia californica*. *Botanical Gazette* 94: 826–828.

Gremillion, K. J. 1998. Changing roles of wild and cultivated plant resources among early farmers of eastern Kentucky. *Southeastern Archaeology* 17: 140–157.

Gugerli, F. 2008. Old seeds coming in from the cold. *Science* 322: 1789–1790.

Haak, D. C., L. A. McGinnis, D. J. Levey, and J. J. Tewksbury. 2011. Why are not all chilies hot? A trade-off limits pungency. *Proceedings of the Royal Society B* 279: 2012–2017.

Hanson, T. R., S. J. Brunsfeld, and B. Finegan. 2006. Variation in seedling density and seed predation indicators for the emergent tree *Dipteryx panamensis* in continuous and fragmented rainforest. *Biotropica* 38: 770–774.

Hanson, T. R., S. J. Brunsfeld, B. Finegan, and L. P. Waits. 2007. Conventional and genetic measures of seed dispersal for *Dipteryx panamensis* (Fabaceae) in continuous and fragmented Costa Rican rainforest. *Journal of Tropical Ecology* 23: 635–642.

———. 2008. Pollen dispersal and genetic structure of the tropical tree *Dipteryx panamensis* in a fragmented landscape. *Molecular Ecology* 17: 2060–2073.

Harden, B. 1996. *A River Lost: The Life and Death of the Columbia*. New York: W. W. Norton.

Hargrove, J. L. 2006. History of the calorie in nutrition. *Journal of Nutrition* 136: 2957–2961.

———. 2007. Does the history of food energy units suggest a solution to "Calorie confusion"? *Nutrition Journal* 6: 44.

Hart, K. 2002. *Eating in the Dark: America's Experiment with Genetically Engineered Food*. New York: Pantheon Books.

Haufler, C. H. 2008. Species and speciation. In T. A. Ranker and C. H. Haufler, eds., *Biology and Evolution of Ferns and Lyophytes*. Cambridge, UK: Cambridge University Press.

Henig, R. M. 2000. *The Monk in the Garden*. Boston: Houghton Mifflin.

Heraclitus. 2001. *Fragments*. New York: Penguin.

Herrera, C. M. 1989. Seed dispersal by animals: A role in angiosperm diversification? *American Naturalist* 133: 309–322.

Herrera, C. M., and O. Pellmyr. 2002. *Plant-Animal Interactions: An Evolutionary Approach.* Oxford: Blackwell Sciences.

Hewavitharange, P., S. Karunaratne, and N. S. Kumar. 1999. Effect of caffeine on shot-hole borer beetle *Xyleborus fornicatus* of tea *Camellia sinensis. Phytochemistry* 51: 35–41.

Hillman, G., R. Hedges, A. Moore, S. College, et al. 2001. New evidence of Late glacial cereal cultivation at Abu Hureyra on the Euphrates. *Holocene* 11: 383–393.

Hirschel, E. H., H. Prem, and G. Madelung. 2004. *Aeronautical Research in Germany—From Lilienthal Until Today.* Berlin: Springer-Verlag.

Hollingsworth, R. G., J. W. Armstrong, and E. Campbell. 2002. Caffeine as a repellent for slugs and snails. *Nature* 417: 915–916.

Hooker, J. D. 1847. An enumeration of the plants of the Galapagos Archipelago; with descriptions of those which are new. *Transactions of the Linnean Society of London, Botany* 20: 163–233.

———. 1847. On the vegetation of the Galapagos Archipelago, as compared with that of some other tropical islands and of the continent of America. *Transactions of the Linnean Society of London, Botany* 20: 235–262.

Huffman, M. 2001. Self-medicative behavior in the African great apes: An evolutionary perspective into the origins of human traditional medicine. *BioScience* 51: 651–661.

Iltis, H. 1966. *Life of Mendel.* Reprint of 1932 translation by E. and C. Paul. New York: Hafner.

Janzen, D. H., and P. S. Martin. 1982. Neotropical anachronisms: The fruits the gomphotheres ate. *Science* 215: 19–27.

Jolly, C. J. 1970. The seed-eaters: A new model of hominid differentiation based on a baboon analogy. *Man* 5: 5–26.

Kadman, L. 1957. A coin find at Masada. *Israel Exploration Journal* 7: 61–65.

Kahn, V. 1987. Characterization of starch isolated from avocado seeds. *Journal of Food Science* 52: 1646–1648.

Kalugin, O. 2009. *Spymaster: My Thirty-Two Years in Intelligence and Espionage Against the West.* New York: Basic Books.

Kardong, K., and V. L. Bels. 1998. Rattlesnake strike behavior: Kinematics. *Journal of Experimental Biology* 201: 837–850.

Kingsbury, J. M. 1992. Christopher Columbus as a botanist. *Arnoldia* 52: 11–28.

Kingsbury, N. 2009. *Hybrid: The History and Science of Plant Breeding.* Chicago: University of Chicago Press.

Klauber, L. M. 1956. *Rattlesnakes: Their Habits, Life Histories, and Influence on Mankind,* vols. 1 and 2. Berkley: University of California Press.

Klein, H. S. 2002. The structure of the Atlantic slave trade in the 19th century: An assessment. *Outre-mers* 89: 63–77.

Knight, M. H. 1995. Tsamma melons: *Citrullus lanatus,* a supplementary water supply for wildlife in the southern Kalahari. *African Journal of Ecology* 33: 71–80.

Koltunow, A. M., T. Hidaka, and S. P. Robinson. 1996. Polyembry in citrus. *Plant Physiology* 110: 599–609.

Krauss, R. 1945. *The Carrot Seed.* New York: HarperCollins.

Lack, D. 1947. *Darwin's Finches.* Cambridge, UK: Cambridge University Press.

Le Couteur, P., and J. Burreson. 2003. *Napoleon's Buttons: 17 Molecules That Changed History.* New York: Jeremy P. Tarcher / Penguin.

Lee, H. 1887. *The Vegetable Lamb of Tartary.* London: Sampson Low, Marsten, Searle and Rivington.

Lee-Thorp, J., A. Likius, H. T. Mackaye, P. Vignaud, et al. 2012. Isotopic evidence for an early shift to C4 resources by Pliocene hominins in Chad. *Proceedings of the National Academy of Sciences* 109: 20369–20372.

Lemay, S., and J. T. Hannibal. 2002. *Trigonocarpus excrescens* Janssen 1940, a supposed seed from the Pennsylvanian of Illinois, is a millipede (Diplopida: Euphoberiidae). *Kirtlandia* 53: 37–40.

Levey, D. J., J. J. Tewksbury, M. L. Cipollini, and T. A. Carlo. 2006. A Weld test of the directed deterrence hypothesis in two species of wild chili. *Oecologica* 150: 51–68.

Levin, D. A. 1990. Seed banks as a source of genetic novelty in plants. *American Naturalist* 135: 563–572.

Lev-Yadun, S. 2009. Aposematic (warning) coloration in plants. Pp. 167–202 in F. Baluska, ed., *Plant-Environment Interactions: Signaling and Communication in Plants.* Berlin: Springer-Verlag.

Lim, M. 2012. Clicks, cabs, and coffee houses: Social media and oppositional movements in Egypt, 2004–2011. *Journal of Communication* 62: 231–248.

Lobova, T., C. Geiselman, and S. Mori. 2009. *Seed Dispersal by Bats in the Neotropics.* New York: New York Botanical Garden.

Loewer, P. 1995. *Seeds: The Definitive Guide to Growing, History & Lore.* Portland, OR: Timber Press.

Loskutov, I. G. 1999. *Vavilov and His Institute: A History of the World Collection of Plant Genetic Resources in Russia.* Rome: International Plant Genetic Resources Institute.

Lucas. P., P. Constantino, B. Wood, and B. Lawn. 2008. Dental enamel as a dietary indicator in mammals. *BioEssays* 30: 374–385.

Lucas, P. W., J. T. Gaskins, T. K. Lowrey, M. E. Harrison, et al. 2011. Evolutionary optimization of material properties of a tropical seed. *Journal of the Royal Society Interface* 9: 34–42.

Machnicki, N. J. 2013. How the chili got its spice: Ecological and evolutionary interactions between fungal fruit pathogens and wild chilies. Ph.D. dissertation, University of Washington, Seattle.

Mannetti, L. 2011. Understanding plant resource use by the ≠Khomani Bushmen of the southern Kalahari. Master's thesis, University of Stellenbosch, South Africa.

Martins, V. F., P. R. Guimaraes Jr., C. R. B. Haddad, and J. Semir. 2009. The effect of ants on the seed dispersal cycle of the typical myrmecochorous *Ricinus communis. Plant Ecology* 205: 213–222.

Marwat, S. K., M. J. Khan, M. A. Khan, M. Ahmad, et al. 2009. Fruit plant species mentioned in the Holy Quran and Ahadith and their ethnomedicinal importance. *American-Eurasian Journal of Agricultural and Environmental Sciences* 5: 284–295.

Masi, S., E. Gustafsson, M. Saint Jalme, V. Narat, et al. 2012. Unusual feeding behavior in wild great apes, a window to understand origins of self-medication in humans: Role of sociality and physiology on learning process. *Physiology and Behavior* 105: 337–349.

McLellan, D., ed. 2000. *Karl Marx: Selected Writings.* Oxford: Oxford University Press.

Mendel, G. 1866. Experiments in plant hybridization. Translated by W. Bateson and R. Blumberg. *Verhandlungen des naturforschenden Vereines in Brünn, Bd. IV für das Jahr 1865,* Abhandlungen: 3–47.

Mercader, J. 2009. Mozambican grass seed consumption during the Middle Stone Age. *Science* 326: 1680–1683.

Mercader, J., T. Bennett, and M. Raja. 2008. Middle Stone Age starch acquisition in the Niassa Rift, Mozambique. *Quaternary Research* 70: 283–300.

Mercier, S. 1999. The evolution of world grain trade. *Review of Agricultural Economics* 21: 225–236.

Midgley, J. J., K. Gallaher, and L. M. Kruger. 2012. The role of the elephant (*Loxodonta africana*) and the tree squirrel (*Paraxerus cepapi*) in marula (*Sclerocarya birrea*) seed predation, dispersal and germination. *Journal of Tropical Ecology* 28: 227–231.

Moore, A. M. T., G. C. Hillman, and A. J. Legge. 2000. *Village on the Euphrates: From Foraging to Farming at Abu Hureyra*. Oxford: Oxford University Press.

Moseley, C. W. R. D, trans. 1983. *The Travels of Sir John Mandeville*. London: Penguin.

Murray, D. R., ed. 1986. *Seed Dispersal*. Orlando, FL: Academic Press.

Nathan, R., F. M. Schurr, O. Spiegel, O. Steinitz, et al. 2008. Mechanisms of long-distance seed dispersal. *Trends in Ecology and Evolution* 23: 638–647.

Nathanson, J. A. 1984. Caffeine and related methylxanthines: Possible naturally occurring pesticides. *Science* 226: 184–187.

Newman, D. J., and G. M. Cragg. 2012. Natural products as sources of new drugs over the 30 years from 1981 to 2010. *Journal of Natural Products* 75: 311–335.

Peterson, K., and M. E. Reed. 1994. *Controversy, Conflict, and Compromise: A History of the Lower Snake River Development*. Walla Walla, WA: US Army Corps of Engineers, Walla Walla District.

Piperno, D. R., E. Weiss, I. Holst, and D. Nadel. 2004. Processing of wild cereal grains in the Upper Paleolithic revealed by starch grain analysis. *Nature* 430: 670–673.

Pollan, M. 2001. *The Botany of Desire*. New York: Random House.

Porter, D. M. 1980. Charles Darwin's plant collections from the voyage of the *Beagle*. *Journal of the Society for the Bibliography of Natural History* 9: 515–525.

———. 1984. Relationships of the Galapagos flora. *Biological Journal of the Linnean Society* 21: 243–251.

Preedy, V. R., R. R. Watson, and V. B. Patel. 2011. *Nuts and Seeds in Health and Disease*. London: Academic Press.

Pringle, P. 2008. *The Murder of Nikolai Vavilov*. New York: Simon and Schuster.

Ramsbottom, J. 1942. Recent work on germination. *Nature* 149: 658.

Ranker, T. A., and C. H. Haufler, eds. 2008. *Biology and Evolution of Ferns and Lyophytes*. Cambridge, UK: Cambridge University Press.

Raven, P. H., R. F. Evert, and S. E. Eichhorn. 1992. *Biology of Plants*, 5th ed. New York: Worth Publishers.

Reddy, S. N. 2009. Harvesting the landscape: Defining protohistoric plant exploitation in coastal Southern California. SCA *Proceedings* 22: 1–10.

Rettalack, G. J., and D. L. Dilcher. 1988. Reconstructions of selected seed ferns. *Annals of the Missouri Botanical Garden* 75: 1010–1057.

Riello, G. 2013. *Cotton: The Fabric That Made the Modern World.* Cambridge, UK: Cambridge University Press.

Rosentiel, T. N., E. E. Shortlidge, A. N. Melnychenko, J. F. Pankow, et al. 2012. Sex-specific volatile compounds influence microarthropod-mediated fertilization of moss. *Nature* 489: 431–433.

Rothwell, G. W., and R. A. Stockey. 2008. Phylogeny and evolution of ferns: A paleontological perspective. Pp. 332–366 in T. A. Ranker and C. H. Haufler, eds., *Biology and Evolution of Ferns and Lyophytes.* Cambridge, UK: Cambridge University Press.

Sacks, O. 2008. *Musicophilia.* New York: Vintage.

Sallon, S., E. Solowey, Y. Cohen, R. Korchinsky, et al. 2008. Germination, genetics, and growth of an ancient date seed. *Science* 320: 1464.

Sathakopoulos, D. C. 2004. *Famine and Pestilence in the Late Roman and Early Byzantine Empire.* Birmingham Byzantine and Ottoman Monographs, vol. 9. Aldershot Hants, UK: Ashgate.

Scharpf, R. F. 1970. Seed viability, germination, and radicle growth of dwarf mistletoe in California. USDA Forest Service Research Paper PSW-59. Berkeley, CA: Pacific SW Forest and Range Experiment Station.

Schivelbusch, W. 1992. *Tastes of Paradise: A Social History of Spices, Stimulants, and Intoxicants.* New York: Pantheon Books.

Schopfer, P. 2006. Biomechanics of plant growth. *American Journal of Botany* 93: 1415–1425.

Scotland, R. W., and A. H. Wortley. 2003. How many species of seed plants are there? *Taxon* 52: 101–104.

Seabrook, J. 2007. Sowing for the apocalypse: The quest for a global seed bank. *New Yorker,* August 7, 60–71.

Sharif, M. 1948. Nutritional requirements of flea larvae, and their bearing on the specific distribution and host preferences of the three Indian species of *Xenopsylla* (Siphonaptera). *Parasitology* 38: 253–263.

Shaw, George Bernard. 1918. The vegetarian diet according to Shaw. Reprinted in *Vegetarian Times,* March/April 1979, 50–51.

Sheffield, E. 2008. Alteration of generations. Pp. 49–74 in T. A. Ranker and C. H. Haufler, eds., *Biology and Evolution of Ferns and Lyophytes.* Cambridge, UK: Cambridge University Press.

Shen-Miller, J., J. William Schopf, G. Harbottle, R. Cao, et al. 2002. Long-living lotus: Germination and soil γ-irradiation of centuries-old fruits, and cultivation, growth, and phenotypic abnormalities of offspring. *American Journal of Botany* 89: 236–247.

Simpson, B. B., and M. C. Ogorzaly. 2001. *Economic Botany*, 3rd ed. Boston: McGraw Hill.

Stephens, S. G. 1958. Salt water tolerance of seeds of *Gossypium* species as a possible factor in seed dispersal. *American Naturalist* 92: 83–92.

———. 1966. The potentiality for long range oceanic dispersal of cotton seeds. *American Naturalist* 100: 199–210.

Stöcklin, J. 2009. Darwin and the plants of the Galápagos Islands. *Bauhinia* 21: 33–48.

Strait, D. S., P. Constantino, P. Lucas, B. G. Richmond, et al. 2013. Viewpoints: Diet and dietary adaptations in early hominins. The hard food perspective. *American Journal of Physical Anthropology* 151: 339–355.

Strait, D. S., G. W. Webe, S. Neubauer, J. Chalk, et al. 2009. The feeding biomechanics and dietary ecology of *Australopithecus africanus*. *Proceedings of the National Academy of Sciences* 106: 2124–2129.

Swan, L. W. 1992. The Aeolian biome. *BioScience* 42: 262–270.

Taviani, P. E., C. Varela, J. Gil, and M. Conti. 1992. *Christopher Columbus: Accounts and Letters of the Second, Third, and Fourth Voyages*. Rome: Instituto Poligrafico e Zecca Dello Stato.

Telewski, F. W., and J. D. Zeevaart. 2002. The 120-year period for Dr. Beal's seed viability experiment. *American Journal of Botany* 89: 1285–1288.

Tewksbury, J. J., D. J. Levey, M. Huizinga, D. C. Haak, et al. 2008. Costs and benefits of capsaicin-mediated control of gut retention in dispersers of wild chilies. *Ecology* 89: 107–117.

Tewksbury, J. J., and G. P. Nabhan. 2001. Directed deterrence by capsaicin in chilies. *Nature* 412: 403–404.

Tewksbury, J. J., G. P. Nabhan, D. Norman, H. Suzan, et al. 1999. In situ conservation of wild chiles and their biotic associates. *Conservation Biology* 13: 98–107.

Tewksbury, J. J., K. M. Reagan, N. J. Machnicki, T. A. Carlo, et al. 2008. Evolutionary ecology of pungency in wild chilies. *Proceedings of the National Academy of Sciences* 105: 11808–11811.

Theophrastus. 1916. *Enquiry into Plants and Minor Works on Odours and Weather Signs*, vol. 2. Translated by A. Hort. New York: G. P. Putnam's Sons.

Thompson, K. 1987. Seeds and seed banks. *New Phytologist* 26: 23–34.

Thoresby, R. 1830. *The Diary of Ralph Thoresby, F.R.S.* London: Henry Colburn and Richard Bentley.

Tiffney, B. 2004. Vertebrate dispersal of seed plants through time. *Annual Review of Ecology, Evolution, and Systematics* 35: 1–29.

Traveset, A. 1998. Effect of seed passage through vertebrate frugivores' guts on germination: A review. *Perspectives in Plant Ecology, Evolution and Systematics* 1/2: 151–190.

Tristram, H. B. 1865. *The Land of Israel: A Journal of Travels in Palestine.* London: Society for Promoting Christian Knowledge.

Turner, J. 2004. *Spice: The History of a Temptation.* New York: Vintage.

Ukers, W. H. 1922. *All About Coffee: A History of Coffee from the Classic Tribute to the World's Most Beloved Beverage.* New York: Tea and Coffee Trade Journal Company.

Unger, R. W. 2004. *Beer in the Middle Ages and the Renaissance.* Philadelphia: University of Pennsylvania Press.

United States Bureau of Reclamation. 2000. *Horsetooth Reservoir Safety of Dam Activities—Final Environmental Impacts Assessment, EC-1300–00–02.* Loveland, CO: United States Bureau of Reclamation, Eastern Colorado Area Office.

Valster, A. H., and P. K. Hepler. 1997. Caffeine inhibition of cytokinesis: Effect on the phragmoplast cytoskeleton in living *Tradescantia* stamen hair cells. *Protoplasma* 196: 155–166.

Vander Wall, S. B. 2001. The evolutionary ecology of nut dispersal. *Botanical Review* 67: 74–117.

Van Wyhe, J., ed. 2002. The Complete Work of Charles Darwin Online, http://darwin-online.org.uk.

Vozzo, J. A., ed. 2002. *Tropical Tree Seed Manual.* Agriculture Handbook 721. Washington, DC: United States Department of Agriculture Forest Service.

Walters, D. R. 2011. *Plant Defense: Warding off Attack by Pathogens, Herbivores, and Parasitic Plants.* Oxford: Wiley-Blackwell.

Walters, R. A., L. R. Gurley, and R. A. Toby. 1974. Effects of caffeine on radiation-induced phenomena associated with cell-cycle traverse of mammalian cells. *Biophysical Journal* 14: 99–118.

Weckel, M., W. Giuliano, and S. Silver. 2006. Jaguar (*Panthera onca*) feeding ecology: Distribution of predator and prey through time and space. *Journal of Zoology* 270: 25–30.

Weiner, J. 1995. *The Beak of the Finch: A Story of Evolution in Our Time.* New York: Alfred A. Knopf.

Wendel, J. F., C. L. Brubaker, and T. Seelanan. 2010. The origin and evolution of *Gossypium.* Pp. 1–18 in J. M. Stewart, et al., eds., *Physiology of Cotton.* Dordrecht, Netherlands: Springer.

Whealy, D. O. 2011. *Gathering: Memoir of a Seed Saver.* Decorah, IA: Seed Savers Exchange.

Whiley, A. W., B. Schaffer, and B. N. Wolstenholme. 2002. *The Avocado: Botany, Production and Uses.* Cambridge, MA: CABI Publishing.

Whitney, K. 2002. Dispersal for distance? *Acacia ligulata* seeds and meat ants *Iridomyrmex viridiaeneus. Austral Ecology* 27: 589–595.

Willis, K. J., and J. C. McElwain. 2002. *The Evolution of Plants.* Oxford: Oxford University Press.

Willson, M. 1993. Mammals as seed-dispersal mutualists in North America. *Oikos* 67: 159–167.

Wing, L. D., and I. O. Buss. 1970. Elephants and forests. *Wildlife Monographs* 19: 3–92.

Woodburn, J. H. 1999. *20th Century Bioscience: Professor O. J. Eigsti and the Seedless Watermelon.* Raleigh, NC: Pentland Press.

Wrangham, R. W. 2009. *Catching Fire: How Cooking Made Us Human.* New York: Basic Books.

———. 2011. Honey and fire in human evolution. Pp. 149–167 in J. Sept and D. Pilbeam, eds., *Casting the Net Wide: Papers in Honor of Glynn Isaac and His Approach to Human Origins Research.* Oxford: Oxbow Books.

Wrangham, R. W., and R. Carmody. 2010. Human adaptation to the control of fire. *Evolutionary Anthropology* 19: 187–199.

Wright, G. A., D. D. Baker, M. J. Palmer, D. Stabler, et al. 2013. Caffeine in floral nectar enhances a pollinator's memory of reward. *Science* 339: 1202–1204.

Yadin, Y. 1966. *Masada: Herod's Fortress and the Zealots' Last Stand.* New York: Random House.

Yafa, S. 2005. *Cotton: The Biography of a Revolutionary Fiber.* New York: Penguin.

Yarnell, R. A. 1978. Domestication of sunflower and sumpweed in eastern North America. Pp. 289–300 in R. I. Ford, ed., *The Nature and Status of Ethnobotany.* Anthropological Papers No. 67. Ann Arbor: University of Michigan Museum of Anthropology.

Yashina, S., S. Gubin, S. Maksimovich, A. Yashina, et al. 2012. Regeneration of whole fertile plants from 30,000-y-old fruit tissue buried in Siberian permafrost. *Proceedings of the National Academy of Sciences* 109: 4008–4013.

Young, F. 1906. *Christopher Columbus and the New World of His Discovery*. London: E. Grant Richards.

Zacks, R. 2002. *The Pirate Hunter: The True Story of Captain Kidd*. New York: Hyperion.

Index

Thor Hanson is a conservation biologist, Guggenheim Fellow, Switzer Environmental Fellow, and winner of the John Burroughs Medal. The author of *Feathers* and *The Impenetrable Forest*, he lives with his wife and son on an island in Washington State.